Charles Hatwell Horne Cheyne

The Earth's Motion of Rotation

Including the Theory of Precession and Nutation

Charles Hatwell Horne Cheyne

The Earth's Motion of Rotation
Including the Theory of Precession and Nutation

ISBN/EAN: 9783337002541

Printed in Europe, USA, Canada, Australia, Japan

Cover: Foto ©berggeist007 / pixelio.de

More available books at **www.hansebooks.com**

THE EARTH'S MOTION

OF ROTATION

INCLUDING THE THEORY OF

PRECESSION AND NUTATION.

BY

C. H. H. CHEYNE, M.A.

ST JOHN'S COLLEGE, CAMBRIDGE,
SECOND MATHEMATICAL MASTER IN WESTMINSTER SCHOOL,
AUTHOR OF "AN ELEMENTARY TREATISE ON THE PLANETARY THEORY."

London and Cambridge:
MACMILLAN AND CO.
1867.

PREFACE.

In offering to the student a treatment of the Problem of the Earth's rotation somewhat different from that which has been usually given in elementary text-books, a few words of explanation are necessary.

The FIRST PART consists of an application of the method of the variation of elements to the general problem of rotation. That the formulæ for calculating these elements are identical in the motions both of translation and rotation, appeared so remarkable, that it might be well to present the latter in a form easily accessible. As far as I am aware, an elementary investigation of these formulæ has not yet been given: in attempting to supply this, I have adopted a method somewhat similar to that which I have given for the corresponding equations of the motion of translation, in an *Elementary Treatise on the Planetary Theory*. The striking analogy, thus developed, between the solutions of problems, in appearance so dissimilar, may, I hope, lead to a more complete study of Lagrange's beautiful theory of the variation of arbitrary constants.

In the SECOND PART the general rotation-formulæ are applied to the particular case of the Earth. These formulæ afford a simple and accurate proof of the important theorem of the Stability of the Earth's axis and of the motion about it, so far as these depend upon the attractions of distant bodies. In this I have followed M. de Pontécoulant. The remaining pages are devoted to a consideration of the motion of the Earth's axis in space. In this I have obtained the formulæ for calculating Precession and Nutation, first, by an application of the general method, and afterwards, by an independent process; but I have not carried the approximation further than has been usual in elementary text-books.

C. H. H. CHEYNE.

1, Dean's Yard,
    Westminster,
        September, 1867.

CONTENTS.

PART I.

GENERAL THEORY OF ROTATION.

ART.   PAGE
1. Object of this treatise   1
2. Reason of the smallness of the effect upon the rotation of a
    planet of the attractions of other bodies ....   „

Undisturbed Motion.

3. Integration of equations of motion   2
4. Signification of the constants of integration   3
5. Equations for determining the motion in space   4
6. The same when the invariable plane is the plane of reference   6
7. When the motion with reference to the invariable plane is
    known, to determine it with reference to any other plane   8

Disturbed Motion.

8. Method of treating the problem. Elements of the motion.
    Definition of *plane of maximum areas*   9
9. Expression for the sum of the moments of the disturbing
    forces about any line through the centre of gravity of the
    planet   10
10. Various ways of expressing the function V   11
11. Equation for calculating h   13
12. Analogy of this equation to that for calculating the mean
    distance in the Planetary Theory   14
13. Equations of motion ....   „
14. Comparison of these equations with those employed in
    Chapter II. of the *Planetary Theory*   16
15. Equations for calculating a, γ and k   17

ART.   PAGE

16. Second method of obtaining the equation for calculating γ ...   19
17. Relation between certain partial differential coefficients of l,
    ψ_1 and ϕ_1   20
18. Relation between partial differential coefficients of V   22
19. Equation for calculating l   24
20. Equation for calculating g   25
21. Recapitulation of formulæ for calculating the elements   27
22. Comparison of these with the corresponding equations of the
    Planetary Theory ....   „

PART II.

APPLICATION OF PRECEDING RESULTS.
PRECESSION AND NUTATION.

23. Application to the case of the Earth : method of treatment...   31
24. Stability of the Earth's axis, the disturbing force being
    neglected   32
25. The same, including the first power of the disturbing force...   33
26. Stability of the velocity   35
27. Effect of tidal friction ....   „
28. Poisson's equations for calculating θ and ψ   37
29. Expression for V ....   „
30. Form assumed when $B = A$ 39
31. Moment of the disturbing couple due to the Sun's attraction   „
32. Expressions for θ and ψ obtained from Poisson's formulæ   40
33, 34. The same obtained independently   42
35. Examination of these formulæ. Solar Precession and Nuta-
    tion   45
36. Effect of the Moon's action on the motion of the Earth's axis
    with reference to the plane of the Moon's orbit ....   „
37. The same with reference to the ecliptic : values of θ and ψ...   47
38. Luni-solar Precession   50
39. Geometrical representation of the motion of the Earth's axis
    in consequence of Precession and Nutation   51
40. Annual Luni-solar Precession   52
41. Effect of Planetary attraction inappreciable ....   „

THE EARTH'S MOTION OF ROTATION.

PART I.

GENERAL THEORY OF ROTATION.

1. In that part of Physical Astronomy which usually goes by the name of the Planetary Theory we are concerned with the motions of translation only of the planets in space : we now propose to consider their motions of rotation. The principles of the conservation of the motions of translation and rotation permit us to consider these separately, and to treat the latter as if the centre of gravity were a fixed point. We shall adopt a method perfectly rigorous, and free from all assumptions, with the single exception of the hypothesis, already required in the Planetary Theory, that the attracting bodies are so distant that their action may be supposed the same as it would be if their whole masses were condensed into their centres of gravity. Thus we shall obtain, for the determination of the motion, formulæ applicable to the case of any planet or other rigid body : an interesting application of these will then be afforded by the special circumstances which occur in the *Earth's Motion of Rotation.*

2. If the planets were exactly spherical in shape, it is clear that the attractions of the Sun, Moon, and of the other planets could produce no effect upon their rotation, since they would all pass through the centre of gravity. But

C.

although this is not the case, yet the deviation from exact sphericity being very small, the motion will differ only slightly from what it would be if these disturbing forces did not exist. We shall, therefore, by neglecting them obtain first an approximate solution of the problem, and then by the method of the variation of parameters deduce from it accurate results. Since, however, the motion of a rigid body about a fixed point under the action of no forces is discussed in works on Rigid Dynamics, we shall here consider it only so far as is necessary for the purpose of obtaining results which will be required in the sequel.

Undisturbed Motion.

3. Let ω_1, ω_2, ω_3 be the angular velocities of a planet about the principal axes at its centre of gravity; A, B, C the moments of inertia about these axes: then Euler's equations give

$$A\frac{d\omega_1}{dt} - (B - C)\,\omega_2\omega_3 = 0,$$

$$B\frac{d\omega_2}{dt} - (C - A)\,\omega_3\omega_1 = 0,$$

$$C\frac{d\omega_3}{dt} - (A - B)\,\omega_1\omega_2 = 0.$$

Multiplying these equations by ω_1, ω_2, ω_3 respectively, adding, and integrating, we have

$$A\omega_1^2 + B\omega_2^2 + C\omega_3^2 = h,$$

where h is the constant of integration.

Again, multiplying by $A\omega_1$, $B\omega_2$, $C\omega_3$, adding, and integrating, we have

$$A^2\omega_1^2 + B^2\omega_2^2 + C^2\omega_3^2 = k^2,$$

where k^2 is the constant of integration.

From these two equations we obtain

$$\omega_1^2 = \frac{k^2 - Bh + (B - C)\,C\omega_3^2}{(A - B)\,A},$$

$$\omega_2^2 = \frac{-k^2 + Ah + (C - A)\,C\omega_3^2}{(A - B)\,B}.$$

Substituting these values, the third equation of motion becomes

$$C\frac{d\omega_3}{dt} = \frac{\{k^2 - Bh + (B - C)\,C\omega_3^2\}^{\frac{1}{2}}\{-k^2 + Ah + (C - A)\,C\omega_3^2\}^{\frac{1}{2}}}{\surd(AB)},$$

whence $t + l$

$$= C\surd(AB)\int\frac{d\omega_3}{\{k^2 - Bh + (B - C)\,C\omega_3^2\}^{\frac{1}{2}}\{-k^2 + Ah + (C - A)\,C\omega_3^2\}^{\frac{1}{2}}},$$

where l is the constant of integration.

This integral cannot in the general case be found; we may however approximate: thus t is known in terms of ω_3, and consequently ω_3 in terms of t; and then from above, ω_1, ω_2 are also known.

4. With respect to the constants introduced by the integration, we may remark that h represents the *vis viva* (Routh's *Rigid Dynamics*, Art. 194), and k the area conserved on the invariable plane. To prove the latter point, the areas conserved on the principal planes being $A\omega_1$, $B\omega_2$, $C\omega_3$ (Routh's *R. D.*, Art 179), and the direction cosines of the invariable plane with reference to the principal axes

$$\frac{A\omega_1}{k}, \quad \frac{B\omega_2}{k}, \quad \frac{C\omega_3}{k}$$

(Routh's *R. D.*, Art. 125), the area conserved on the invariable plane

$$= \frac{A^2\omega_1^2}{k} + \frac{B^2\omega_2^2}{k} + \frac{C^2\omega_3^2}{k}$$

$$= \frac{A^2\omega_1^2 + B^2\omega_2^2 + C^2\omega_3^2}{k} = k \text{ (Art. 3).}$$

5. When ω_1, ω_2, ω_3 are known at any time, the resultant angular velocity of the planet is known, and also the position of the instantaneous axis of rotation with reference to the principal axes. It remains to shew how the position of these axes in space may be determined.

Suppose a sphere described with its centre at the centre of gravity of the planet and its radius of any magnitude: take as a plane of reference any fixed plane passing through the centre of gravity, and let it cut the sphere in the great circle ON; also let the principal plane of xy cut the sphere in the great circle NAB, N being the node of this plane upon the fixed plane, and A, B, C the points where the sphere is cut by the principal axes of x, y, z. Take P the pole of ON, and join PA, PB, CA, CB by arcs of great circles.

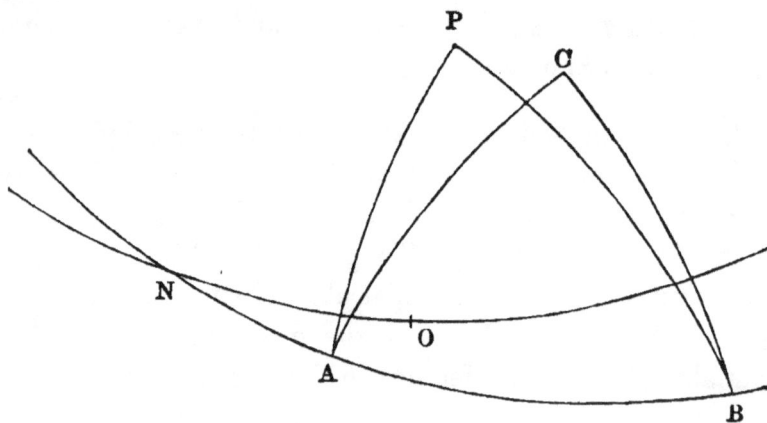

Let the angle $ONA = \theta$, $ON = \psi$, $NA = \phi$: then if the angles θ, ϕ, ψ be known, the position of the planet will be determined with reference to the fixed plane.

Now we may consider the planet to be moving with angular velocities ω_1, ω_2, ω_3 about the principal axes; or with angular velocities $\dfrac{d\psi}{dt}$, $\dfrac{d\theta}{dt}$, $\dfrac{d\phi}{dt}$, the first about a normal to the fixed plane, the second about the line of nodes, the third about the principal axis of z. We shall adopt the usual convention with respect to signs, and consider positive those angular velocities which tend to turn the planet round the axes of x, y, z from y to z, z to x, x to y, respectively.

Thus, resolving about the principal axes, we have

$$\omega_1 = -\frac{d\theta}{dt}\cos\phi - \frac{d\psi}{dt}\cos PA,$$

$$\omega_2 = \frac{d\theta}{dt}\sin\phi - \frac{d\psi}{dt}\cos PB,$$

$$\omega_3 = \frac{d\phi}{dt} - \frac{d\psi}{dt}\cos\theta.$$

Now, by Spherical Trigonometry,

$$\cos PA = -\sin\theta\sin\phi,$$

$$\cos PB = -\sin\theta\cos\phi;$$

therefore

$$\omega_1 = -\frac{d\theta}{dt}\cos\phi + \frac{d\psi}{dt}\sin\theta\sin\phi,$$

$$\omega_2 = \frac{d\theta}{dt}\sin\phi + \frac{d\psi}{dt}\sin\theta\cos\phi,$$

$$\omega_3 = \frac{d\phi}{dt} - \frac{d\psi}{dt}\cos\theta.$$

Hence also,

$$\frac{d\psi}{dt} \sin \theta = \omega_1 \sin \phi + \omega_2 \cos \phi,$$

$$\left.\begin{array}{c} \frac{d\theta}{dt} = -\omega_1 \cos \phi + \omega_2 \sin \phi, \\[2mm] \frac{d\phi}{dt} - \frac{d\psi}{dt} \cos \theta = \omega_3, \end{array}\right\} \quad \ldots\ldots\text{(A)}^*.$$

By substituting the values of ω_1, ω_2, ω_3 obtained as above (Art 3), and then integrating these equations, θ, ϕ, and ψ would be determined, and thus the position of the principal axes at any time would be known. The integration, however, cann in general be effected; so that we are obliged to have recourse to a special hypothesis with regard to the position of the fixed plane of reference. If we take for this purpose the invariable plane, the process becomes much simplified.

6. Let then θ_1, ϕ_1, ψ_1 denote relatively to the invariabl plane the same angles which relatively to the original plane of reference have been denoted by θ, ϕ, ψ. Then, the direction cosines of the invariable plane with reference to the principal axes being

$$\frac{A\omega_1}{k}, \quad \frac{B\omega_2}{k}, \quad \frac{C\omega_3}{k}$$

respectively, we have (see figure of preceding Article)

$$\frac{A\omega_1}{k} = \cos PA = -\sin \theta_1 \sin \phi_1,$$

$$\frac{B\omega_2}{k} = \cos PB = -\sin \theta_1 \cos \phi_1,$$

$$\frac{C\omega_3}{k} = \cos \theta_1;$$

---

* There is much disagreement between writers as to the measurement of the angles employed in these kinematical equations; the above, however, agrees with La Place, Poisson, and Pontécoulant.

therefore

$$\tan \phi_1 = \frac{A\omega_1}{B\omega_2} = \sqrt{\left\{ \frac{A}{B} \cdot \frac{k^2 - Bh + (B - C)\, C\omega_3^2}{-k^2 + Ah + (C - A)\, C\omega_3^2} \right\}} \quad \text{by Art. 3,}$$

$$\cos \theta_1 = \frac{C\omega_3}{k}.$$

These equations give θ_1 and ϕ_1: to obtain ψ_1, substitute in the first of the equations (A) of Art. 5; thus

$$\frac{d\psi_1}{dt} \sin^2 \theta_1 = \omega_1 \sin \theta_1 \sin \phi_1 + \omega_2 \sin \theta_1 \cos \phi_1;$$

therefore

$$\frac{d\psi_1}{dt}(k^2 - C^2\omega_3^2) = -(A\omega_1^2 + B\omega_2^2)\, k$$

$$= -(h - C\omega_3^2)\, k;$$

therefore

$$\frac{d\psi_1}{dt} = -\frac{h - C\omega_3^2}{k^2 - C^2\omega_3^2}\, k;$$

combining this with the result of Art. 3, and integrating, we have

$$\psi_1 + g = -kC\sqrt{(AB)} \times$$

$$\int \frac{(h - C\omega_3^2)\, d\omega_3}{(k^2 - C^2\omega_3^2)\{h^2 - Bh + (B - C)C\omega_3^2\}^{\frac{1}{2}}\{-k^2 + Ah + (C - A)C\omega_3^2\}^{\frac{1}{2}}},$$

where g is the constant of integration.

Since ω_3 is known in terms of t from Art. 3, these equations give θ_1, ϕ_1, ψ_1; so that the position of the principal axes is known at any time with reference to the invariable plane. Since, however, when the disturbing forces are taken into account, this plane ceases to be absolutely invariable, it will be convenient to be able to refer the motion to some other plane which does remain fixed, and which may be taken as a plane of reference: this we can now do by Spherical Trigonometry.

7. Let the surface of a sphere of any radius, with its centre at the centre of gravity of the planet, be cut by the fixed plane of reference, the invariable plane, and the principal plane of xy, in the great circles OMN, MI, INA respectively.

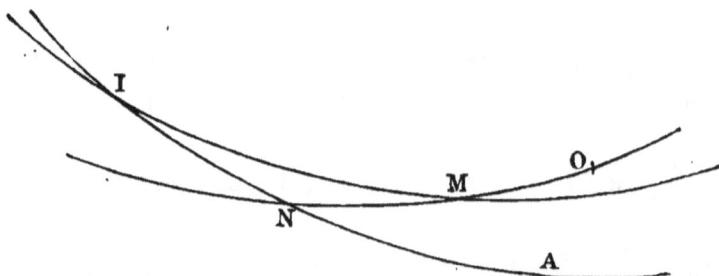

As before let $ON = \psi$, $NA = \phi$, the angle $ONA = \theta$; also take M as the origin from which ψ_1 is measured, and let $MI = \psi_1$, $IA = \phi_1$, the angle $MIN = \theta_1$: let OM (the longitude of the node of the invariable plane) $= \alpha$, and IMN (its inclination to the plane of reference) $= \gamma$.

Then the sides of the spherical triangle IMN are

$$\psi - \alpha, \quad \psi_1, \quad \phi_1 - \phi;$$

and the angles respectively opposite to these,

$$\theta_1, \quad \pi - \theta, \quad \gamma.$$

Hence, by the formulæ of Spherical Trigonometry, we have

$$\cos \theta = \cos \gamma \cos \theta_1 - \sin \gamma \sin \theta_1 \cos \psi_1,$$
$$\sin (\phi_1 - \phi) \sin \theta = \sin \gamma \sin \psi_1,$$
$$\sin (\psi - \alpha) \sin \theta = \sin \theta_1 \sin \psi_1,$$

which determine θ, ϕ, ψ, when θ_1, ϕ_1, ψ_1, α, γ are known.

Disturbed Motion.

8. Having now shewn how to determine the position and velocity of rotation of the planet on the hypothesis that no forces act upon it, we proceed to a rigorous treatment of the problem. We shall employ the principle of the variation of parameters, and suppose the results already obtained to represent the true solution, the arbitrary constants or elements being no longer constants, but variable quantities, which it will be our object to determine. We shall speak of the forces which produce this variation as *disturbing forces*.

The elements which have been already employed in the undisturbed motion are six in number, viz. h, k, l, g, a, γ: in considering these as variable we shall arrive at the very remarkable result that the equations for calculating their variations are precisely the same as the corresponding equations for the motion of translation in the Planetary Theory.

DEF. The plane of which the direction cosines are

$$\frac{A\omega_1}{k}, \quad \frac{B\omega_2}{k}, \quad \frac{C\omega_3}{k},$$

the area conserved upon which has been shewn (Art. 4) to be equal to k, will in future be termed the *plane of maximum areas*, on account of the property which it possesses, that k is a maximum[*]; since it can no longer be considered invariable.

---

[*] See Routh's *Rigid Dynamics*, Art. 174.

9. *To find an expression for the sum of the moments of the disturbing forces about any line through the centre of gravity of the planet.*

We shall suppose the disturbing body (which may be the Sun, Moon, or another planet), so distant that it may be considered to attract as if condensed into its centre of gravity.

Let m be the mass of the disturbed, m' of the disturbing body, ρ_1 the distance of the centre of gravity of the latter from an element δm_1 of the former : also let $V_1 = \dfrac{m'}{\rho_1}$. Then, if σ denote the length of the arc of any curve measured from some fixed point to the element δm_1, the disturbing force on this element in direction of the tangent to the curve, and tending to increase σ, will be

$$\delta m_1 \frac{dV_1}{d\sigma}.$$

If we suppose this arbitrary curve so drawn that its tangent at the point where the element is situated is perpendicular to the axis about which the moments are to be taken, and denote by p the distance of δm_1 from this axis, the moment of the force will be

$$\delta m_1 . p \frac{dV_1}{d\sigma}.$$

Let the small arc $\delta\sigma$ subtend an angle $\delta\chi$ at the nearest point of the axis; then $\delta\sigma = p\,\delta\chi$, and the moment becomes

$$\delta m_1 \frac{dV_1}{d\chi}.$$

Similarly, if V_2 refer to an element δm_2, the moment of the disturbing force on this element will be

$$\delta m_2 \frac{dV_2}{d\chi},$$

$\delta\chi$ being the same as for the element δm_1 since these elements are supposed rigidly connected.

Hence the sum of the moments of the disturbing forces on all the elements of the planet

$$= \delta m_1 \frac{dV_1}{d\chi} + \delta m_2 \frac{dV_2}{d\chi} + \dots$$

$$= \frac{d}{d\chi} \Sigma (\delta m_1 . V_1) ;$$

or, if we write V' for $\Sigma (\delta m_1 . V_1)$, the sum of the moments will be $$\frac{dV'}{d\chi},$$

where $$V' = m'\Sigma \frac{\delta m}{\rho}.$$

Cor. If there are several disturbing bodies m', m'', &c. and V', V'', &c. are the functions corresponding, the sum of the moments of the forces due to their action will still be $\frac{dV}{d\chi}$, where $V = V' + V'' + \dots$

The result of this Article may be thus enunciated:—*Suppose a small arbitrary rotation given to the planet about any axis through an angle $\delta\chi$; then, supposing V expressed in terms of χ and quantities which do not vary in this hypothetical motion, the sum of the moments of the disturbing forces about this axis will be expressed by the partial differential coefficient* $\frac{dV}{d\chi}$.

The function V is thus clearly analogous to the disturbing function R of the Planetary Theory.

10. We may express V in various ways which will be found useful:

(i) As a function of θ, ϕ, ψ. Let x, y, z be the coordinates of an element δm of the disturbed planet, the fixed plane of reference being that of xy, and the axis of x the line from which ψ and α are measured; let x', y', z' be the co-ordinates of the centre of gravity of the disturbing body referred to the same axes; and let x_1, y_1, z_1 be the co-ordinates of δm referred to the principal axes of the planet. Then

$$V = m'\Sigma\frac{\delta m}{\rho}, \quad \text{(Art. 9)},$$

$$= m'\Sigma\frac{\delta m}{\sqrt{(x'-x)^2 + (y'-y)^2 + (z'-z)^2}}.$$

Now if λ, μ, ν be the angles which the fixed axis of x makes with the principal axes of x, y, z, we have

$$x = x_1\cos\lambda + y_1\cos\mu + z_1\cos\nu;$$

and by Spherical Trigonometry (see fig. to Art. 5),

$$\cos\lambda = \cos\phi\cos\psi + \sin\phi\sin\psi\cos\theta,$$

$$\cos\mu = -\sin\phi\cos\psi + \cos\phi\sin\psi\cos\theta,$$

$$\cos\nu = \sin\psi\sin\theta.$$

Thus x may be expressed in terms of θ, ϕ, ψ and of the co-ordinates of δm referred to the principal axes. Similarly, y and z may be expressed in terms of θ, ϕ, ψ. If the values of x, y, z so obtained be now substituted in the expression for V, it will become a function of θ, ϕ, ψ; and, in so far as it depends upon the disturbed planet, of quantities independent of the time.

(ii) As a function of θ_1, ϕ_1, ψ_1, α, γ. By Art. 7, we have,

$$\theta = f(\theta_1, \psi_1, \gamma),$$

$$\phi - \phi_1 = f(\theta_1, \psi_1, \gamma),$$

$$\psi - \alpha = f(\theta_1, \psi_1, \gamma),$$

the symbol f denoting a different function in each case. If then we suppose. V to have been expressed as a function of θ, ϕ, ψ by (i), these equations will enable us to express it as a function of θ_1, ϕ_1, ψ_1, α, γ.

(iii) As a function of t and the elements. Collecting together the results of Arts. 6 and 3, and making $C\omega_3 = s$, we may write

$$\theta_1 = f(k, s),$$

$$\phi_1 = f(h, k, s),$$

$$\psi_1 + g = f(h, k, s),$$

$$t + l = f(h, k, s).$$

Supposing V to have been expressed as a function of θ_1, ϕ_1, ψ_1, α, γ, by (ii), the first three of these equations will enable us to express it as a function of s, h, k, g, α, γ: then, if s be eliminated by means of the fourth equation, it will become a function of $t + l$, h, k, g, α, γ; that is, of t and the elements.

11. We now proceed to obtain equations for calculating the values of the elements at any time, commencing with h, the element of *vis viva*.

Let T denote the *vis viva* due to the rotation of the planet, $\delta m_1 \dfrac{dV_1}{d\sigma}$ the resolved part of the disturbing force on an element δm_1 of its mass in the direction of motion of the element: then the equation of *vis viva* gives

$$\frac{dT}{dt} = 2\Sigma\left(\delta m_1 \frac{dV_1}{d\sigma}\frac{d\sigma}{dt}\right)$$

$$= 2\Sigma\left(\delta m_1 \frac{d(V_1)}{dt}\right),$$

where in $\dfrac{d(V_1)}{dt}$, the differential coefficient is taken only in so far as t is involved through the co-ordinates of the element δm_1 of the disturbed planet.   This equation may be written

$$\frac{dT}{dt} = 2\frac{d}{dt}\,\Sigma\,(\delta m_1.\,V_1)$$

$$= 2\,\frac{d(V)}{dt}\,,$$

if, as in Art. 9, we write $V = \Sigma\,(\delta m_1.\,V_1)$.   Now from the result of (iii) in the preceding Article we notice that $t + l$ always occurs in V as one quantity; therefore

$$\frac{d(V)}{dt} = \frac{dV}{dl}\,;$$

also in the undisturbed motion $T = h$; hence our equation becomes

$$\frac{dh}{dt} = 2\,\frac{dV}{dl}\,;$$

from which h may be calculated.

12.   This equation may also be written

$$\frac{dh}{dt} = 2\,\frac{d(V)}{dt}\,,$$

under which form it is easily seen to be identical with the corresponding equation

$$\frac{d}{dt}\left(-\frac{\mu}{a}\right) = 2\,\frac{d(R)}{dt}\,,$$

of the Planetary Theory.   (See *Planetary Theory*, Art. 26.) In fact it appears that the element of *vis viva* will be given by a similar equation in all cases of motion, whether of translation or rotation, when the disturbing bodies attract according to any law expressed by a function of the distance.

Equations of Motion.

13. Let k_1, k_2, k_3 denote the areas conserved about three rectangular axes of x, y, z, originating in the centre of gravity of the disturbed planet, and moving with angular velocities β_1, β_2, β_3 about their instantaneous positions; also let L, M, N be the moments of the disturbing forces about these axes: then (Routh's *Rigid Dynamics*, Art. 120) we have the equations of motion

$$\frac{dk_1}{dt} - k_2\beta_3 + k_3\beta_2 = L,$$

$$\frac{dk_2}{dt} - k_3\beta_1 + k_1\beta_3 = M,$$

$$\frac{dk_3}{dt} - k_1\beta_2 + k_2\beta_1 = N.$$

Now let k_3 denote the area conserved on the plane of maximum areas: then from the undisturbed motion (Art. 4) we have

$$k_3 = k;$$

and since the areas conserved upon all planes perpendicular to the plane of maximum areas are zero*, we must also have

$$k_1 = 0, \quad k_2 = 0,$$

always; and therefore

$$\frac{dk_1}{dt} = 0, \quad \frac{dk_2}{dt} = 0:$$

---

* This follows from the fact that the area conserved upon any plane is equal to the projection upon it of the area conserved upon the plane of maximum areas. (Routh's *Rigid Dynamics*, Art. 174.)

hence our equations become

$$k\beta_2 = L,$$

$$-k\beta_1 = M,$$

$$\frac{dk}{dt} = N.$$

We may express β_1 and β_2 in terms of $\dfrac{d\alpha}{dt}$ and $\dfrac{d\gamma}{dt}$, α and γ being respectively the longitude of the node and the inclination of the plane of maximum areas to the fixed plane of reference. Since no assumption has yet been made with regard to the position of the axis of x, it will be convenient to suppose it to coincide with the line of nodes. Employing the equations of Art. 5, and making $\omega_1 = \beta_1$, $\omega_2 = \beta_2$, $\theta = \gamma$, $\phi = 0$, $\psi = \alpha$, we have

$$\frac{d\alpha}{dt} \sin \gamma = \beta_2,$$

$$\frac{d\gamma}{dt} = -\beta_1,$$

and the equations of motion become

$$k\frac{d\alpha}{dt} \sin \gamma = L,$$

$$k\frac{d\gamma}{dt} = M,$$

$$\frac{dk}{dt} = N.$$

14.   With respect to the last of these equations, we may remark that it is of the same form as it would be if the plane of maximum areas were still invariable. Now on referring to Art. 18 of the *Planetary Theory* it will be seen that the equations for the motion of translation, when referred to axes in the plane of the orbit, take the same forms

as if this plane were at rest, and moreover, that the second equation there obtained is identical with the third of the preceding Article*. This coincidence is not accidental, but arises from the fact that in the motion of translation, the plane of maximum areas *is* the plane of the orbit. In fact, the equations obtained in the preceding Article for the motion of rotation are equally applicable to the motion of translation, and we shall see that the resulting formulæ for the calculation of the elements involved are identical in the two motions.

15. It now only remains to find expressions for L, M, N in terms of the differential coefficients of the function V, in order to obtain from the equations of Art. 13 formulæ for calculating the elements α, γ and k. We shall suppose V expressed as a function of θ_1, ϕ_1, ψ_1, α, γ (Art. 10 (ii)).

To find L:—Suppose a small arbitrary rotation given to the planet about the axis of x, that is, about the line of nodes of the plane of maximum areas: then we may represent the change in position of the planet by supposing γ alone to vary, θ_1, ϕ_1, ψ_1, α remaining constant†. Thus, the tendency of L being to diminish γ, we have

$$L = -\frac{dV}{d\gamma},$$

the differential coefficient being partial. Hence by the first equation of motion,

$$\frac{d\alpha}{dt} = -\frac{1}{k \sin \gamma}\frac{dV}{d\gamma}.$$

* The quantity here denoted by k corresponds with what in the Planetary Theory is denoted by h.

† This is equivalent to supposing the plane of maximum areas fixed in the body during the rotation, a supposition perfectly allowable, since the rotation given to the planet is hypothetical, and has nothing to do with its actual

To find M:—Suppose a small arbitrary rotation given to the planet through an angle $\delta\chi$ about the axis of y, that is, about a line through L the centre of gravity of the planet, in the plane of maximum areas perpendicular to LM its line of nodes : let the effect of the rotation be to change the position of this plane from MI to mi: draw Mn perpendicular to mi.

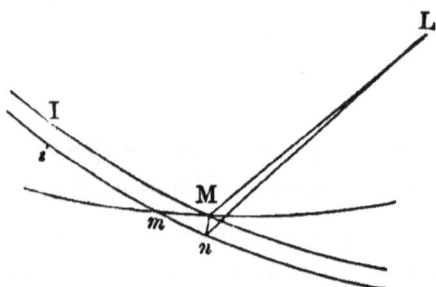

Then $Mn = \delta\chi$, $mn = -\delta\psi_1$, $Mm = \delta\alpha$, the angle $Mmn = \gamma$: and we have $\delta\chi = \delta\alpha \sin\gamma$, $\quad \delta\psi_1 = -\delta\alpha \cos\gamma$;

therefore $\quad \dfrac{d\alpha}{d\chi} = \dfrac{1}{\sin\gamma}$, $\quad \dfrac{d\psi_1}{d\chi} = -\cos\gamma \dfrac{d\alpha}{d\chi} = -\cot\gamma.$

And $\qquad M = \dfrac{dV}{d\chi} = \dfrac{dV}{d\alpha}\dfrac{d\alpha}{d\chi} + \dfrac{dV}{d\psi_1}\dfrac{d\psi_1}{d\chi},$

since we may suppose α and ψ_1 alone to vary with χ; therefore

$$M = \frac{1}{\sin\gamma}\frac{dV}{d\alpha} - \cot\gamma\,\frac{dV}{d\psi_1}.$$

Now (Art. 10 (iii)) $\quad \psi_1 + g = f(h, k, s)$,

and θ_1, ϕ_1 do not involve g; therefore

$$\frac{dV}{d\psi_1} = -\frac{dV}{dg}:$$

motion. Any other supposition might be adopted with regard to the motion of this plane, but this is the most convenient, and will be retained throughout this Article.

hence
$$M = \frac{1}{\sin \gamma} \frac{dV}{d\alpha} + \cot \gamma \frac{dV}{dg};$$

and the second equation of motion gives
$$\frac{d\gamma}{dt} = \frac{1}{k \sin \gamma} \frac{dV}{d\alpha} + \frac{\cos \gamma}{h \sin \gamma} \frac{dV}{dg}.$$

To find N:—Suppose a small arbitrary rotation given to the planet about the axis of z, that is, about a normal through the centre of gravity of the planet to the plane of maximum areas: then we may suppose ψ_1 alone to vary, and since the tendency of N is to diminish ψ_1, we shall have

$$N = -\frac{dV}{d\psi_1} = \frac{dV}{dg},$$

as in the preceding Article; then the third equation gives

$$\frac{dk}{dt} = \frac{dV}{dg}.$$

16. The equation for calculating γ may also be simply obtained as follows: if we refer the motion to the fixed plane of reference, and suppose V expressed as a function of θ, ϕ, ψ, since $k \cos \gamma$ is the area conserved on this plane, we shall have

$$\frac{d}{dt}(k \cos \gamma) = -\frac{dV}{d\psi}.$$

Now (Art. 10 (ii)), $\psi - \alpha = f(\theta_1, \psi_1, \gamma)$, and θ, ϕ do not involve α; therefore

$$\frac{dV}{d\psi} = \frac{dV}{d\alpha},$$

the latter differential coefficient supposing V expressed as a function either of θ_1, ϕ_1, ψ_1, α, γ, or simply of t and the elements. Hence

$$\frac{d}{dt}(k \cos \gamma) = -\frac{dV}{d\alpha};$$

therefore $\qquad k \sin \gamma \dfrac{d\gamma}{dt} = \dfrac{dV}{da} + \cos \gamma \dfrac{dk}{dt}$

$$= \dfrac{dV}{da} + \cos \gamma \dfrac{dV}{dg};$$

therefore $\qquad \dfrac{d\gamma}{dt} = \dfrac{1}{k \sin \gamma} \dfrac{dV}{da} + \dfrac{\cos \gamma}{k \sin \gamma} \dfrac{dV}{dg}.$

The method of this Article is identical with the second method given in Art. 34 of the *Planetary Theory* for the determination of $\dfrac{di}{dt}$; we reserve the comparison of the results to a subsequent Article (Art. 22).

17.   The following proposition will be useful in finding $\dfrac{dl}{dt}$.

To shew that $\dfrac{dl}{dk} = 2\dfrac{d\psi_1}{db}$, *and* $\dfrac{dl}{ds} = -2\dfrac{d\phi_1}{dh}$, *the variables being connected together by the equations of Art.* 10, (iii).

Let $\qquad u = k^2 - Bh + \dfrac{B-C}{C}s^2 \left.\begin{array}{c} \\ \\ \\ \\ \end{array}\right\}$ .... (1),

$$v = -k^2 + Ah + \dfrac{C-A}{C}s^2$$

where $s = C\omega_3$.   Then from Arts. 3 and 6, we have

$$t + l = \sqrt{(AB)} \int \dfrac{ds}{\sqrt{(uv)}} \qquad \text{.................... (2)},$$

$$\psi_1 + g = -\sqrt{(AB)} \int \dfrac{k}{k^2 - s^2} \cdot \dfrac{h - \dfrac{s^2}{C}}{\sqrt{(uv)}} ds \text{ ......(3)},$$

$$\tan^2 \phi_1 = \dfrac{Au}{Bv} \qquad \text{.............................. (4)}.$$

From equation (2), by differentiating under the integral sign,

$$\frac{dl}{dk} = -\frac{1}{2}\sqrt{(AB)}\int \frac{u\dfrac{dv}{dk} + v\dfrac{du}{dk}}{(uv)^{\frac{3}{2}}}\, ds$$

$$= \sqrt{(AB)}\int \frac{k\,(u-v)}{(uv)^{\frac{3}{2}}}\, ds, \text{ by equations (1).}$$

From equation (3), in like manner, we have

$$\frac{d\psi_1}{dh} = -\sqrt{(AB)}\int \frac{k}{k^2-s^2}\cdot \frac{uv - \dfrac{1}{2}\left(h - \dfrac{s^2}{C}\right)\left(u\dfrac{dv}{dh} + v\dfrac{du}{dh}\right)}{(uv)^{\frac{3}{2}}}\, ds$$

$$= -\frac{1}{2}\sqrt{(AB)}\int \frac{k}{(uv)^{\frac{3}{2}}}\cdot \frac{2uv - \left(h - \dfrac{s^2}{C}\right)(Au - Bv)}{k^2-s^2}\, ds.$$

To simplify this expression, we find from equations (1),

$$h - \frac{s^2}{C} = \frac{u+v}{A-B},$$

$$k^2 - s^2 = \frac{Au+Bv}{A-B}.$$

Substituting these values, and reducing,

$$\frac{d\psi_1}{dh} = \frac{1}{2}\sqrt{(AB)}\int \frac{k\,(u-v)}{(uv)^{\frac{3}{2}}}\, ds$$

$$= \frac{1}{2}\frac{dl}{dk}, \text{ from above;}$$

therefore
$$\frac{dl}{dk} = 2\frac{d\psi_1}{dh}.$$

Again, from equation (2), we have

$$\frac{dl}{ds} = \sqrt{\left(\frac{AB}{uv}\right)}.$$

Differentiating equation (4), and substituting for $\frac{du}{dh}$ and $\frac{dv}{dh}$ from equations (1),

$$2 \tan \phi_1 \sec^2 \phi_1 \frac{d\phi_1}{dh} = \frac{A}{B} \cdot \frac{-Bv - Au}{v^2} \; ;$$

also from equation (4),

$$\tan \phi_1 \sec^2 \phi_1 = \sqrt{\left(\frac{Au}{Bv}\right)} \cdot \frac{Au + Bv}{Bv} \; ;$$

hence, by substitution,

$$\frac{d\phi_1}{dh} = -\frac{1}{2} \sqrt{\left(\frac{AB}{uv}\right)} \; ;$$

therefore

$$\frac{dl}{ds} = -2 \frac{d\phi_1}{dh} .$$

18. The following relations between partial differential coefficients of the function V, as well as the results of the preceding Article, will be required in finding $\frac{dl}{dt}$.

In Art. 10 we have shewn that, in order to express V as a function of t and the elements, we may first express it as a function of θ_1, ϕ_1, ψ_1, α, γ, by case (ii), and then eliminate θ_1, ϕ_1, ψ_1 by means of the equations in (iii). The first three of these equations are

$$\theta_1 = f(k, s),$$
$$\phi_1 = f(h, k, s),$$
$$\psi_1 + g = f(h, k, s).$$

If, then, we substitute for θ_1, ϕ_1, ψ_1 from these equations, V will become a function of h, k, s, g, α, γ; and we shall have

$$\left(\frac{dV}{dh}\right) = \frac{dV}{d\theta_1}\frac{d\theta_1}{dh} + \frac{dV}{d\phi_1}\frac{d\phi_1}{dh} + \frac{dV}{d\psi_1}\frac{d\psi_1}{dh}$$

$$= \frac{dV}{d\phi_1}\frac{d\phi_1}{dh} + \frac{dV}{d\psi_1}\frac{d\psi_1}{dh},$$

since θ_1 does not involve h. We have used the brackets to distinguish the partial differential coefficient with respect to h, of V expressed as here supposed, from its partial differential coefficient when expressed as a function of t and the elements. In order to pass from the expression of V as a function of h, k, s, g, α, γ to its expression as a function of t and the elements, we must solve the fourth equation of Art. 10 (iii), with respect to s; thus we shall have

$$s = f(h, k, t+l),$$

and then

$$\frac{dV}{dh} = \left(\frac{dV}{dh}\right) + \frac{dV}{ds}\frac{ds}{dh};$$

or, writing for $\left(\dfrac{dV}{dh}\right)$ the value obtained above,

$$\frac{dV}{dh} = \frac{dV}{ds}\frac{ds}{dh} + \frac{dV}{d\phi_1}\frac{d\phi_1}{dh} + \frac{dV}{d\psi}\frac{d\psi_1}{dh},$$

a relation which will be required in the following Article.

Again, differentiating the equation

$$s = f(h, k, t+l)$$

on the hypothesis that h and l alone vary, we have

$$0 = \frac{ds}{dh} + \frac{ds}{dl}\frac{dl}{dh},$$

where $\dfrac{dl}{dh}$ is the same as would be obtained by direct differ-entiation of the last equation of Art. 10 (iii).   Hence

$$\frac{dV}{ds}\frac{ds}{dh} = -\frac{dV}{ds}\frac{ds}{dl}\frac{dl}{dh}$$

$$= -\frac{dV}{dl}\frac{dl}{dh},$$

since l is introduced into V only through s. ⸱ This relation also will be required in the following Article.

19.   We are now in a position to find $\dfrac{dl}{dt}$, and shall for this purpose refer the motion to the principal plane of xy. We must first, however, obtain an expression for the sum of the moments of the disturbing forces about the axis of z. Consider, then, V as a function of θ_1, ϕ_1, ψ_1, α, γ, and let a small arbitrary displacement be given to the planet about the axis of z.   We may express this by supposing θ_1, ψ_1, α, γ to remain constant while ϕ_1 alone varies.   Since, then, with the usual convention with respect to signs, the ten-dency of the moment we are considering is to increase ϕ_1, it will be expressed by $\dfrac{dV}{d\phi_1}$.   Thus, from Euler's equations of motion,

$$C\frac{d\omega_3}{dt} - (A - B)\,\omega_1\omega_2 = \frac{dV}{d\phi_1};$$

whence, writing as before, s for $C\omega_3$,

$$\frac{ds}{dt} = (A - B)\,\omega_1\omega_2 + \frac{dV}{d\phi_1}.$$

From the results obtained in the case of undisturbed motion, we have (Art. 10 (iii))

$$t + l = f(h, k, s).$$

If we differentiate this equation, first considering the elements variable, and then considering them invariable, and equate the 'results, substituting the above value of $\frac{ds}{dt}$, all terms not depending upon the disturbing force will disappear: thus

$$\frac{dl}{dt} = \frac{dl}{dh}\frac{dh}{dt} + \frac{dl}{dk}\frac{dk}{dt} + \frac{dl}{ds}\frac{dV}{d\phi_1};$$

or, substituting the values of $\frac{dh}{dt}$ and $\frac{dk}{dt}$ from Arts. 11 and 15,

$$\frac{dl}{dt} = 2\frac{dV}{dl}\frac{dl}{dh} + \frac{dV}{dg}\frac{dl}{dk} + \frac{dV}{d\phi_1}\frac{dl}{ds}:$$

but by Art. 17,

$$\frac{dl}{dk} = 2\frac{d\psi_1}{dh}, \quad \frac{dl}{ds} = -2\frac{d\phi_1}{dh},$$

also, as in Art. 15,

$$\frac{dV}{dg} = -\frac{dV}{d\psi_1},$$

and by Art. 18,

$$\frac{dV}{dl}\frac{dl}{dh} = -\frac{dV}{ds}\frac{ds}{dh};$$

therefore $\frac{dl}{dt} = -2\left(\frac{dV}{ds}\frac{ds}{dh} + \frac{dV}{d\psi_1}\frac{d\psi_1}{dh} + \frac{dV}{d\phi_1}\frac{d\phi_1}{dh}\right)$

$$= -2\frac{dV}{dh} \text{ (Art. 18)},$$

from which l may be calculated.

20. To obtain the formula for g, we might of course proceed in the same manner, differentiating the third equation of Art. 10 (iii); but the process would be complicated by the fact that ψ_1 is measured from a moving point and on

a moving plane, and consequently that the form of $\dfrac{d\psi_{\prime}}{dt}$ is not the same in the disturbed as in the undisturbed motion. We shall therefore employ a different method: availing ourselves of the formulæ already obtained for calculating five out of the six elements, we shall obtain the sixth by direct substitution, in the same manner as that in which the epoch was obtained in Art. 36 of the *Planetary Theory*.

By Art. 10 (iii), we may write

$$V = f(t + l,\ g,\ h,\ k,\ a,\ \gamma).$$

Differentiating, the elements being considered variable,

$$\frac{dV}{dt} = \frac{dV}{d(t+l)}\left(1 + \frac{dl}{dt}\right) + \frac{dV}{dg}\frac{dg}{dt} + \frac{dV}{dh}\frac{dh}{dt} + \frac{dV}{dk}\frac{dk}{dt}$$

$$+ \frac{dV}{da}\frac{da}{dt} + \frac{dV}{d\gamma}\frac{d\gamma}{dt}:$$

differentiating as if the elements were invariable,

$$\frac{dV}{dt} = \frac{dV}{d(t+l)}.$$

Equating the two values of $\dfrac{dV}{dt}$, we have

$$\frac{dV}{dl}\frac{dl}{dt} + \frac{dV}{dg}\frac{dg}{dt} + \frac{dV}{dh}\frac{dh}{dt} + \frac{dV}{dk}\frac{dk}{dt}$$

$$+ \frac{dV}{da}\frac{da}{dt} + \frac{dV}{d\gamma}\frac{d\gamma}{dt} = 0.$$

Substituting in this equation the values of $\dfrac{dl}{dt}$, $\dfrac{dh}{dt}$, &c., obtained in the preceding Articles, we find after reduction

$$\frac{dg}{dt} = -\frac{dV}{dk} - \frac{\cos\gamma}{k\sin\gamma}\frac{dV}{d\gamma},$$

from which g may be calculated.

21. We will here recapitulate the formulæ which have been obtained for calculating the elements of the motion :

(i) $\dfrac{dh}{dt} = 2\dfrac{dV}{dl}$,

(ii) $\dfrac{dl}{dt} = -2\dfrac{dV}{dh}$,

(iii) $\dfrac{dk}{dt} = \dfrac{dV}{dg}$,

(iv) $\dfrac{dg}{dt} = -\dfrac{dV}{dk} - \dfrac{\cos\gamma}{k\sin\gamma}\dfrac{dV}{d\gamma}$,

(v) $\dfrac{d\alpha}{dt} = -\dfrac{1}{k\sin\gamma}\dfrac{dV}{d\gamma}$,

(vi) $\dfrac{d\gamma}{dt} = \dfrac{1}{k\sin\gamma}\dfrac{dV}{d\alpha} + \dfrac{\cos\gamma}{k\sin\gamma}\dfrac{dV}{dg}$.

22. These six equations agree with those obtained by Pontécoulant in his *Théorie Analytique du Système du Monde* (Tome II. p. 189). In Art. 12 we have shewn that the first of them can be expressed in a form in which it is seen to be identical with the formula from which the mean distance is calculated in the Planetary Theory: it will be instructive to compare the remaining equations with those of the motion of translation. In order to do this, we shall first express the latter in terms of elements having a significa-tion analogous to that of the above. Now in the motion of rotation h is the element of *vis viva*, and this in the motion of translation is equal to $-\dfrac{\mu}{a}$: k is the area conserved on the plane of maximum areas, which plane in the motion of translation becomes that of the orbit, and the area conserved upon it is equal to $\sqrt{\mu a(1-e)^2}$: l is the element added to t

in the integration, and this in the motion of translation is $\frac{1}{n}(\epsilon - \varpi)$: g is the element added to ψ_1, and this corresponds to ϖ in the motion of translation, only we must remember that g is measured wholly on the plane of maximum areas from its node, while ϖ is measured on the fixed plane of reference as far as the node, and thence on the plane of the orbit; so that $\varpi - \Omega$ will correspond to g: lastly, α is the longitude of the node of the plane of maximum areas, and γ its inclination to the plane of reference; in the motion of translation the same quantities defining the position of the plane of the orbit are denoted by Ω and i. In order, then, to compare the rotation formulæ with those of translation, we proceed to replace the elements a, e, ϵ, ϖ of the Planetary Theory by four new elements h, k, l, g, connected with the former by the relations

$$h = -\frac{\mu}{a},$$

$$k^2 = \mu a (1 - e^2),$$

$$l = \frac{1}{n}(\epsilon - \varpi),$$

$$g = \varpi - \Omega.$$

Let R' express the form which the function R takes when the new elements are substituted for the old: then we have

$$\frac{dR}{da} = \frac{dR'}{dh}\frac{dh}{da} + \frac{dR'}{dk}\frac{dk}{da} + \frac{dR'}{dl}\frac{dl}{da}*$$

$$= \frac{\mu}{a^2}\frac{dR'}{dh} + \frac{1}{2}na\sqrt{1-e^2}\frac{dR'}{dk} + \frac{3(\epsilon - \varpi)}{2na}\frac{dR'}{dl},$$

---

* In $\dfrac{dR}{da}$, which occurs only in the formula for the epoch, the differential coefficient is supposed to be taken with respect to a only in so far as a

$$\frac{dR}{de} = \frac{dR'}{dk}\frac{dk}{de} = -\frac{na^2e}{\sqrt{1-e^2}}\frac{dR'}{dk},$$

$$\frac{dR}{d\epsilon} = \frac{dR'}{dl}\frac{dl}{d\epsilon} = \frac{1}{n}\frac{dR'}{dl},$$

$$\frac{dR}{d\varpi} = \frac{dR'}{dl}\frac{dl}{d\varpi} + \frac{dR'}{dg}\frac{dg}{d\varpi}$$

$$= -\frac{1}{n}\frac{dR'}{dl} + \frac{dR'}{dg},$$

$$\frac{dR}{d\Omega} = \frac{dR'}{d\Omega} + \frac{dR'}{dg}\frac{dg}{d\Omega}$$

$$= \frac{dR'}{d\Omega} - \frac{dR'}{dg}.$$

Now if we differentiate the above expressions for h, k, l, g, and then substitute the values of $\dfrac{da}{dt}$, $\dfrac{de}{dt}$, &c. obtained in the Planetary Theory, and those obtained here for $\dfrac{dR}{da}$, $\dfrac{dR}{de}$, &c., they will reduce to

$$\frac{dh}{dt} = 2\frac{dR'}{dl},$$

$$\frac{dl}{dt} = -2\frac{dR'}{dh},$$

$$\frac{dk}{dt} = \frac{dR'}{dg},$$

$$\frac{dg}{dt} = -\frac{dR'}{dh} - \frac{\cos i}{k\sin i}\frac{dR'}{di},$$

occurs explicitly in R: if we suppose a to vary also as contained implicitly in n, this differential coefficient will include the term proportional to the time, which may therefore with this understanding be omitted. In Art. 37 of the *Planetary Theory* this term was removed by a different transformation.

$$\frac{d\Omega}{dt} = \frac{1}{k \sin i} \frac{dR'}{di},$$

$$\frac{di}{dt} = -\frac{1}{k \sin i} \frac{dR'}{d\Omega} + \frac{\cos i}{k \sin i} \frac{dR'}{dg}.$$

On comparing these with the rotation formulæ, it will be seen that they are identical, with the exception only that the sign of α differs from that of Ω; and this is accounted for by the fact that α and Ω are measured in opposite directions. Thus, by employing elements having a like signification in the two motions of translation and rotation, we have arrived at the very remarkable result that *the complete solution of the problem of Planetary perturbation, whether in the motion of translation or of rotation, is expressed by the above simple formulæ.*

PART II.

23. THE formulæ obtained in the *first part* are sufficient completely to determine the motion of a planet or other rigid body about its centre of gravity. They are perfectly rigorous, subject only to the hypothesis that the disturbing bodies may be supposed to attract as if condensed into their respective centres of gravity; a hypothesis admissible if these bodies are either very distant or nearly spherical in form. But though the formulæ are exact, they can be integrated only by approximation. We propose, therefore, in the *second part* to restrict ourselves to the particular case of the Earth, taking advantage of such of the results of observation as may be required to enable us to approximate. In order to treat the problem fully we shall consider, first, the motion of the axis of rotation in the Earth itself, with the velocity of rotation about it; secondly, the motion of this axis in space. The first is of special interest, since any change in the position of the axis in the Earth, were such change possible, would affect the permanence of terrestrial latitudes; any change in the velocity of rotation would affect the length of the day. The second is of great importance to astronomers, since it establishes the fact that the first point of Aries, or vernal equinox, to which they are accustomed to refer celestial longitudes, is not a fixed point.

Stability of the axis of rotation in the Earth and of the velocity about it.

24. We proceed then, first, to consider the motion of the axis of rotation within the Earth, and shall be able to shew that, in so far as it depends upon the attractions of other bodies, this axis can never separate appreciably from the axis of figure, and that the velocity about it must always remain appreciably constant.

We have from Art. 3

$$Aω_1^2 + Bω_2^2 + Cω_3^2 = h,$$
$$A^2ω_1^2 + B^2ω_2^2 + C^2ω_3^2 = k^2.$$

Eliminating $ω_3$ from these equations,

$$A(C-A)ω_1^2 + B(C-B)ω_2^2 = Ch - k^2.$$

Now if C be either the greatest or least principal moment, both the terms of the left-hand member of this equation must always retain the same sign; and if we take C about the Earth's axis of figure it will be the greatest and this sign will be positive. We may therefore write

$$Ch - k^2 = (C-A)(C-B)e^2:$$

thus $A(C-A)ω_1^2 + B(C-B)ω_2^2 = (C-A)(C-B)e^2.$

It follows from this that

$ω_1$ can never be greater than $e\sqrt{\dfrac{C-B}{A}}$,

$ω_2$ can never be greater than $e\sqrt{\dfrac{C-A}{B}}$.

If we neglect the disturbing force, h, k and therefore e are constant, and the value of e will be found by substituting.

the values of ω_1, ω_2 as determined by observation at any given epoch. Now the most delicate observations have hitherto shewn no appreciable separation of the axis of rotation from the axis of figure: hence ω_1, ω_2 are at present inappreciable; thus e is so, and we conclude that, independently of the disturbing force, ω_1, ω_2 must always remain inappreciable.

If in the equations of Art. 6, we write $\omega_1 = 0$, $\omega_2 = 0$, we have also $\theta_1 = 0$; so that, if the disturbing force be neglected, we may consider the plane of maximum areas as coincident with the Earth's equator.

25. Let us now examine the effect of the disturbing force upon the angular velocities ω_1, ω_2. By the preceding Article

$$(C-A)\,(C-B)\,e^2 = Ch - k^2.$$

Differentiating, and substituting the expressions for $\dfrac{dh}{dt}$ and $\dfrac{dk}{dt}$ obtained in the *first part* (Art. 21),

$$(C-A)\,(C-B)\,e\,\frac{de}{dt} = C\frac{dV}{dl} - k\frac{dV}{dy}.$$

If the approximation be carried to the first power of the disturbing force, we may, in calculating the second member of this equation, suppose $\theta_1 = 0$: thus the equations of Art. 7 give

$$\theta = \gamma, \quad \phi = \phi_1 - \psi_1, \quad \psi = \alpha.$$

Also, if n be the mean angular velocity, we may to the same order of approximation suppose $\omega_3 = n$: thus, from the equations (A) of Art. 5, ψ and θ are constant and

$$\frac{d\phi}{dt} = n;$$

C.

3

whence by integration

$$\phi = nt + c,$$

c being the value of ϕ at the epoch from which t is measured.

Again, by Art. 10 (iii),

$$\psi_1 + g = f(h, k, s),$$
$$t + l = f(h, k, s).$$

Hence $\qquad \dfrac{dV}{dl} = \dfrac{d(V)}{dt},$

$$\frac{dV}{dg} = -\frac{dV}{d\psi_1} = \frac{dV}{d\phi} = \frac{1}{n}\frac{d(V)}{dt}.$$

Now (Art. 3) $\quad k^2 = A^2\omega_1^2 + B^2\omega_2^2 + C^2\omega_3^2$

$$= C^2 n^2$$

to the same order of approximation : therefore

$$k = Cn.$$

Hence $\quad C\dfrac{dV}{dl} - k\dfrac{dV}{dg} = C\dfrac{d(V)}{dt} - \dfrac{k}{n}\dfrac{d(V)}{dt} = 0;$

and therefore $\qquad\qquad -\dfrac{de}{dt} = 0 :$

whence it follows that to the first order of the disturbing force e continues absolutely constant.

When the approximation is carried to the order of the square of the disturbing force[*], it is found that e^2 can contain no term proportional to the time, nor any inequality, either secular or periodic, raised to the first order in the process of integration. Hence we may consider e^2 as practically constant, and thus conclude, as in the preceding Article, that ω_1, ω_2 will always remain inappreciable.

---

[*] See Pentécoulant's *Système du Monde*, tome II. p. 218.

It follows, as in the preceding Article, that to the same degree of approximation the plane of maximum areas may be regarded as coincident with the Earth's equator.

26. We have now to examine the effect of the disturbing force upon the velocity of rotation. Since ω_1, ω_2 are inappreciable, this velocity will be equal to ω_3; thus, as in the preceding Article, we have

$$\omega_3 = \frac{k}{C};$$

therefore $\quad \dfrac{d\omega_3}{dt} = \dfrac{1}{C}\dfrac{dk}{dt} = \dfrac{1}{C}\dfrac{dV}{dg}$ (Art. 21)

$$= \frac{1}{Cn}\frac{d(V)}{dt}.$$

Now considering V as a function of θ, ϕ, ψ, if our approximation be carried to the first power of the disturbing force, we have

$$\theta = \gamma, \quad \phi = nt + c, \quad \psi = \alpha:$$

so that V will involve t only under the form $nt + c$. It follows that $\dfrac{d(V)}{dt}$ will consist of a series of terms of the form

$$P \, \genfrac{}{}{0pt}{}{\cos}{\sin} \, (pnt + q),$$

where p is some positive integer. And this term will give rise to inequalities whose period is $\dfrac{2\pi}{pn}$, that is, a day or fraction of a day. Now the most delicate observations have hitherto failed to detect any such inequality: hence we must altogether reject such terms, and conclude that ω_3 will always remain sensibly constant.

27. The preceding Articles prove the stability of the axis of rotation in the Earth and of the velocity about it, so

far as the motion depends upon the attractions of distant
bodies; but there is another disturbing influence to be found
in the friction produced by the tides. The effect of this (see
Thomson's and Tait's *Natural Philosophy*) would be to retard
the Earth's velocity; but whether to an appreciable extent
is yet uncertain. The fact, discovered by Professor Adams,
that a portion of the acceleration of the Moon's mean motion
is yet unaccounted for, led Delaunay to suggest that this portion might be apparent only, and really due to the retarding
effect of tidal friction upon the Earth's velocity. But other
possible explanations have been given, and to pursue the
subject further would be beyond the scope of the present
treatise.

Precession and Nutation.

28.   We have now to consider the motion of the Earth's
axis in space, resulting in the phenomena of Precession and
Nutation.   We have seen that, to the first order of the disturbing force, the plane of maximum areas may be considered as coincident with the Earth's equator, and that
consequently to this order of approximation we may write

$$\theta = \gamma, \quad \psi = \alpha.$$

Hence the motion of the equator may be calculated by
formulæ (v) and (vi) of Art. 21.   Now it may be shewn as in
Art. 25 that

$$\frac{dV}{dg} = \frac{1}{n} \frac{d(V)}{dt},$$

and may therefore be neglected, since (as has been proved in
Art. 26) it can give rise only to terms whose period is a day

or fraction of a day, which are wholly insensible. If, then, we write ψ and θ for α and γ respectively, these formulæ become

$$\frac{d\psi}{dt} = - \frac{1}{k \sin \theta} \frac{dV}{d\theta},$$

$$\frac{d\theta}{dt} = \frac{1}{k \sin \theta} \frac{dV}{d\psi};$$

or, since to the same order of approximation $k = C\omega_3 = Cn$,

$$\frac{d\psi}{dt} = - \frac{1}{Cn \sin \theta} \frac{dV}{d\theta},$$

$$\frac{d\theta}{dt} = \frac{1}{Cn \sin \theta} \frac{dV}{d\psi}.$$

29. These elegant equations, due to Poisson, give ψ and θ, and thus determine the motion of the Earth's axis in space. Before we can integrate them it will be necessary to obtain an expression for V.

Let then m be the mass of the Earth, m' that of the disturbing body which we shall suppose to be the Sun; r' the distance between the centres of gravity of these bodies, r the distance from the Earth's centre of gravity of an element δm of its mass, ρ the distance of the same element from the centre of gravity of the Sun: also let the inclination of r to r' be denoted by v. Then by Art. 9,

$$V = m' \Sigma \frac{\delta m}{\rho}$$

$$= m' \Sigma \frac{\delta m}{(r^2 - 2r\,r\cos v + r'^2)^{\frac{1}{2}}}$$

$$= \frac{m}{r'} \Sigma \left\{ \delta m \left(1 - \frac{2r}{r'} \cos v + \frac{r^2}{r'^2} \right)^{-\frac{1}{2}} \right\}.$$

Now since r' is very large in comparison of r, we shall neglect all powers of $\dfrac{r}{r'}$ above the second: thus

$$V = \frac{m'}{r'} \Sigma \left\{ \delta m \left(1 + \frac{r}{r'} \cos v - \frac{1}{2} \frac{r^2}{r'^2} + \frac{3}{2} \frac{r^2}{r'^2} \cos^2 v \right) \right\}$$

$$= \frac{m'}{r'} \Sigma (\delta m) + \frac{m'}{r'^2} \Sigma (\delta m \cdot r \cos v) - \frac{m'}{2r'^3} \Sigma (\delta m \cdot r^2)$$

$$+ \frac{3m'}{2r'^3} \Sigma (\delta m \cdot r^2 \cos^2 v)$$

$$= \frac{m'}{r'} \Sigma (\delta m) + \frac{m'}{r'^2} \Sigma (\delta m \cdot r \cos v) + \frac{m'}{r'^3} \Sigma (\delta m \cdot r^2)$$

$$- \frac{3m'}{2r'^3} \Sigma (\delta m \cdot r^2 \sin^2 v).$$

Of these terms $\dfrac{m'}{r'} \Sigma (\delta m)$ and $\dfrac{m'}{r'^3} \Sigma (\delta m . r^2)$ may be omitted, since they cannot contain the angles θ, ϕ or ψ, and would not be affected by any arbitrary rotation such as is supposed in Art. 9 : also since r is measured from the Earth's centre of gravity, $\Sigma (\delta m . r \cos v) = 0$. Hence we may write

$$V = - \frac{3m'}{2r'^3} \Sigma (\delta m \cdot r^2 \sin^2 v)$$

$$= - \frac{3m'}{2r'^3} Q,$$

if Q denote the moment of inertia of the Earth about the line joining its centre of gravity with that of the Sun. We see, then, that the effective portion of the disturbing function is proportional to this moment of inertia.

30. We have hitherto made no assumption respecting the Earth's figure. We learn from pendulum observations and geodetic measurements that it is approximately an

oblate spheroid; the moments of inertia about all axes in the plane of the equator are therefore nearly equal, and we may take $B = A$. If, then, δ be the Sun's declination,

$$Q = A \cos^2\delta + C \sin^2\delta :$$

also if n' denote the Sun's mean motion, we have approximately

$$\frac{2\pi}{n'} = \frac{2\pi r'^{\frac{3}{2}}}{\sqrt{m + m'}} ;$$

therefore

$$\frac{m + m'}{r'^3} = n'^2,$$

or, neglecting the Earth's mass in comparison with that of the Sun,

$$\frac{m'}{r'^3} = n'^2.$$

Hence, by substitution,

$$V = -\frac{3n'^2}{2} (A \cos^2\delta + C \sin^2\delta).$$

31. By Art. 9, the moment of the Sun's disturbing force about an equatorial diameter of the Earth perpendicular to the line joining the centres of gravity of the two bodies, and tending to increase δ

$$= \frac{dV}{d\delta} = -\frac{3n'^2}{2} (C - A) \sin 2\delta.$$

It is clear that the moment about any axis perpendicular to this is zero, since an hypothetical rotation about any such axis could produce no change in δ. If, then, the Sun's disturbing force be reduced to a parallel force through the Earth's centre of gravity, and a couple, the tendency of the latter will be to give to the equator a rotation towards the ecliptic about an equatorial diameter perpendicular to the line joining the centres of gravity of the two bodies. The moment of this disturbing couple is

$$\frac{3n'^2}{2}(C-A)\sin 2\delta.$$

32. We shall now return to the equations of Art. 28, and deduce from them the motion of the Earth's axis in space. The results obtained will be of sufficient accuracy for our present purpose if we suppose the ecliptic, to which plane the motion will be referred, to remain fixed.

Let ψ denote the longitude of the first point of Aries measured from some fixed point on the ecliptic, λ the Sun's longitude measured from the same point, θ the obliquity of the ecliptic. Then from the spherical triangle formed by the intersection of the equator, the ecliptic, and a declination circle through the Sun, we have

$$\sin \delta = \sin(\psi - \lambda)\sin\theta\,;$$

therefore

$$\cos\delta\,\frac{d\delta}{d\theta} = \sin(\psi - \lambda)\cos\theta,$$

$$\cos\delta\,\frac{d\delta}{d\psi} = \cos(\psi - \lambda)\sin\theta.$$

If $\psi - \lambda = l$, then l will be the Sun's longitude measured from the first point of Aries in the opposite direction to that in which ψ and λ are measured : thus

$$\cos\delta\,\frac{d\delta}{d\theta} = \sin l \cos\theta, \quad \cos\delta\,\frac{d\delta}{d\psi} = \cos l \sin\theta.$$

Hence
$$\frac{dV}{d\theta} = \frac{dV}{d\delta}\frac{d\delta}{d\theta} = -3n'^2(C-A)\sin\delta\sin l\cos\theta,$$

$$\frac{dV}{d\psi} = \frac{dV}{d\delta}\frac{d\delta}{d\psi} = -3n'^2(C-A)\sin\delta\cos l\sin\theta.$$

Substituting these values in the equations of Art. 28,

$$\frac{d\psi}{dt} = \frac{3n'^2}{n} \cdot \frac{C-A}{C} \sin \delta \sin l \cot \theta$$

$$= \frac{3n'^2}{n} \cdot \frac{C-A}{C} \sin^2 l \cos \theta$$

$$= \frac{3n'^2}{2n} \cdot \frac{C-A}{C} \cos \theta \, (1 - \cos 2l).$$

$$\frac{d\theta}{dt} = -\frac{3n'^2}{n} \cdot \frac{C-A}{C} \sin \delta \cos l$$

$$= -\frac{3n'^2}{n} \cdot \frac{C-A}{C} \sin l \cos l \sin \theta$$

$$= -\frac{3n'^2}{2n} \cdot \frac{C-A}{C} \sin \theta \sin 2l.$$

Let I denote the mean value of the obliquity, then since both $\dfrac{n'}{n}$ and $\dfrac{C-A}{C}$ are very small quantities, we may in the second members of these equations write

$$\theta = I, \quad l = n't :$$

thus
$$\frac{d\psi}{dt} = \frac{3n'^2}{2n} \cdot \frac{C-A}{C} \cos I \, (1 - \cos 2n't),$$

$$\frac{d\theta}{dt} = -\frac{3n'^2}{2n} \cdot \frac{C-A}{C} \sin I \sin 2n't.$$

By integration,

$$\psi = \frac{3n'^2}{2n} \cdot \frac{C-A}{C} \cdot \cos I \cdot t - \frac{3n'}{4n} \cdot \frac{C-A}{C} \cdot \cos I \cdot \sin 2n't,$$

$$\theta = I + \frac{3n'}{4n} \cdot \frac{C-A}{C} \cdot \sin I \cdot \cos 2n't,$$

ψ being measured from the position of the first point of Aries at the epoch from which the time is reckoned.

33. These formulæ determine the motion of the Earth's axis in space. Although the method by which they have been obtained has the advantage of connecting the two problems of translation and rotation, it may be worth while to arrive at them by an independent process.

Since the Earth is nearly an oblate spheroid, we shall, as before, take C for the greatest principal moment, and suppose $B = A$, so that all axes in the plane of the equator are principal axes. If, then, taking the plane of the equator for that of xy, we suppose the axes of x and y to revolve with an angular velocity θ_3 about the axis of z, we have the equations of motion*

$$A\frac{d\omega_1}{dt} - A\omega_2\theta_3 + C\omega_2\omega_3 = L,$$

$$A\frac{d\omega_2}{dt} + A\omega_1\theta_3 - C\omega_1\omega_3 = M,$$

$$C\frac{d\omega_3}{dt} = N.$$

Now if we take for the axis of x the projection of the Sun's radius vector on the plane of the equator, we have

$$L = 0, \quad N = 0,$$

$$M = -\frac{3n'^2}{2}(C - A)\sin 2\delta, \quad \text{(Art. 31);}$$

and since θ_3 is very small and occurs only in connexion with ω_1 and ω_2 which are also very small (Arts. 24 and 25), we may write

$$\theta_3 = n' \text{ its mean value.}$$

---

* See Routh's *Rigid Dynamics*, Art. 107: the equations there given reduce to the above if θ_3 is written for $\omega_3 + \frac{d\chi}{dt}$.

Thus the third equation gives

$$C \frac{d\omega_3}{dt} = 0,$$

whence $\quad\quad \omega_3 = \text{const.} = n \text{ suppose};$

and the first two become

$$A \frac{d\omega_1}{dt} + (Cn - An') \omega_2 = 0,$$

$$A \frac{d\omega_2}{dt} - (Cn - An') \omega_1 = -\frac{3n'^2}{2} (C - A) \sin 2\delta:$$

or, denoting $\dfrac{Cn - An'}{A}$ by p,

$$\frac{d\omega_1}{dt} + p\omega_2 = 0,$$

$$\frac{d\omega_2}{dt} - p\omega_1 = -\frac{3n'^2}{2} \cdot \frac{C - A}{A} \sin 2\delta.$$

Now $\dfrac{C - A}{A}$ is a very small fraction; n', the Sun's mean motion, is very small in comparison of n; and δ varies very slowly; we shall, therefore, in integrating these equations, take no account of its variation. Thus, eliminating ω_1,

$$\frac{d^2\omega_2}{dt^2} + p^2\omega_2 = 0,$$

the integral of which is

$$\omega_2 = A \cos (pt - B),$$

where A and B are constants of integration. This term is independent of the disturbing force; but having for its period $\dfrac{2\pi}{p}$, which does not differ much from a day, it must be rejected, since, as we have already remarked in Art. 26, such terms are altogether insensible. Hence we conclude that

$$\omega_2 = 0 ;$$

and by substitution in the second equation,

$$\omega_1 = \frac{3n'^2}{2p} \cdot \frac{C-A}{A} \sin 2\delta$$

$$= \frac{3n'^2}{2n} \cdot \frac{C-A}{C} \left(1 - \frac{An'}{Cn}\right)^{-1} \sin 2\delta$$

$$= \frac{3n'^2}{2n} \cdot \frac{C-A}{C} \sin 2\delta,$$

powers of n' above the second being neglected. We see, then, that the effect of the Sun's attraction is to cause the Earth's axis to travel in a plane perpendicular to that of the disturbing couple which it produces, and with an angular velocity proportional to its moment. This velocity, though so small as to be insensible to the most delicate observations, yet leads to values of ψ and θ which can by no means be neglected.

34.   We proceed to determine the motion of the axes in space. Substituting in the equations of Art. 5, we have

$$\frac{d\psi}{dt} \sin \theta = \frac{3n'^2}{n} \cdot \frac{C-A}{C} \cdot \sin \delta \cos \delta \sin \phi,$$

$$\frac{d\theta}{dt} = - \frac{3n'^2}{n} \cdot \frac{C-A}{C} \cdot \sin \delta \cos \delta \cos \phi.$$

Now from the spherical triangle formed by the intersection of the equator, the ecliptic and a declination circle through the Sun, it is easily seen that

$$\sin \phi = \tan \delta \cot \theta,$$
$$\sin \delta = \sin l \sin \theta,$$
$$\cos l = \cos \phi \cos \delta.$$

By means of these equations we obtain

$$\frac{d\psi}{dt} = \frac{3n'^2}{n} \cdot \frac{C-A}{C} \cdot \cos\theta \sin^2 l,$$

$$\frac{d\theta}{dt} = -\frac{3n'^2}{2n} \cdot \frac{C-A}{C} \cdot \sin\theta \sin 2l;$$

whence, as in Art. 32,

$$\psi = \frac{3n'^2}{2n} \cdot \frac{C-A}{C} \cos I \cdot t - \frac{3n'}{4n} \cdot \frac{C-A}{C} \cdot \cos I \sin 2n't,$$

$$\theta = I + \frac{3n'}{4n} \cdot \frac{C-A}{C} \sin I \cos 2n't.$$

35. It appears from these formulæ that the motion of the Earth's axis is of two kinds; partly secular, partly periodic. It is convenient to consider these separately. The former affects the equinoxes alone, and, being proportional to the time, indicates uniform motion; this motion is one of regression, since ψ has been measured in a direction contrary to that of the apparent motion of the Sun. Considered with reference to the apparent diurnal motion of the stars, the effect is to place the equinoxes in advance of the position they would occupy if fixed: hence it obtained the name of the *Solar Precession of the Equinoxes*. The latter furnishes corrections both on ψ and θ, which go through all their changes in half a year: it is called the *Solar Nutation of the Earth's axis;* the correction on ψ forming the *Nutation in Longitude*, that on θ the *Nutation in Latitude*.

36. We will now examine the effect produced by the Moon's action on the motion of the Earth's axis. Since the investigations which have been given of the effect of the Sun's disturbing force contain nothing to restrict their generality except the special assumptions made with regard to

small quantities, we shall first consider what modifications are required in order that they may be applied in the case of the Moon.

Let ψ', θ', n'', I' denote relatively to the plane of the Moon's orbit the same quantities which relatively to the ecliptic have been denoted by ψ, θ, n', I; also let m, m'' be the masses of the Earth and Moon : then, as in the case of the Sun, we have approximately,

$$\frac{2\pi}{n''} = \frac{2\pi r''^{\frac{3}{2}}}{\sqrt{m + m''}},$$

but we cannot in this case neglect m, as it is in fact much larger than m'' : retaining it, we have

$$\frac{m''}{r'^3} = \frac{m'' n''^2}{m + m''} = \frac{n''^2}{1 + \lambda},$$

if $m = \lambda m''$. Also n'' is small, though not so small as n' in comparison of n. Hence to determine the motion of the Earth's axis with reference to the plane of the Moon's orbit, we have

$$\psi' = \frac{3n''^2}{2n\,(1 + \lambda)} \cdot \frac{C - A}{C} \cdot \cos I' t$$

$$-\frac{3n''}{4n\,(1 + \lambda)} \cdot \frac{C - A}{C} \cdot I' \sin 2n'' t,$$

$$\theta' = I' + \frac{3n''}{4n\,(1 + \lambda)} \cdot \frac{C - A}{C} \cdot \sin I' \cos 2n'' t.$$

The periodical terms in these equations go through all their values in half a month, and are so small that they are usually neglected. Thus we may consider the inclination of the plane of the Earth's orbit to that of the Moon as constant and equal to its mean value, and the precession as uniform and given by

$$\psi' = \frac{3n''^2}{2n\,(1+\lambda)} \cdot \frac{C-A}{C} \cdot \cos I' . t;$$

whence, the velocity of precession

$$\frac{d\psi'}{dt} = \frac{3n''^2}{2n\,(1+\lambda)} \cdot \frac{C-A}{C} \cdot \cos I'.$$

These results, deduced from the corresponding formulæ in the case of the Sun, suppose the plane of the Moon's orbit fixed. Now the line of nodes moves too rapidly to allow of this hypothesis, but the only effect of considering its motion would be to add to $\dfrac{d\psi'}{dt}$ a term depending upon its velocity and not upon the disturbing force of the Moon upon the Earth. Since our object is to trace the effects of this force only, such terms must be omitted.

37. It now remains to determine the motion of the Earth's axis with reference to the ecliptic.

Let θ be the obliquity of the ecliptic, ψ the longitude of the equinox, α the longitude of the node of the Moon's orbit measured from the same origin as ψ, i the inclination of the Moon's orbit to the ecliptic, I' its inclination to the equator. Then supposing ψ' measured from the node of the Moon's orbit on the ecliptic, we have by Spherical Trigonometry (as in Art. 7)

$$\sin(\psi - \alpha)\sin\theta = \sin I' \sin\psi' \quad\ldots\ldots\ldots\ldots\ldots\ldots (1),$$
$$\cos\theta = \cos i \cos I' - \sin i \sin I' \cos\psi' \quad\ldots\ldots\ldots\ldots (2),$$
$$\cos I' = \cos\theta \cos i + \sin\theta \sin i \cos(\psi - \alpha) \quad\ldots\ldots (3).$$

We must now differentiate these equations in order to express $\dfrac{d\psi}{dt}$ and $\dfrac{d\theta}{dt}$ in terms of $\dfrac{d\psi'}{dt}$. In so doing we shall omit the terms involving $\dfrac{d\alpha}{dt}$ and $\dfrac{di}{dt}$, since these quantities depend not upon the disturbing force, but upon the motion of the Moon's orbit. Thus

$$\cos(\psi - \alpha)\sin\theta\,\dfrac{d\psi}{dt} + \sin(\psi - \alpha)\cos\theta\,\dfrac{d\theta}{dt}$$

$$= \sin I' \cos\psi'\,\dfrac{d\psi'}{dt},$$

$$\sin\theta\,\dfrac{d\theta}{dt} = -\sin i \sin I' \sin\psi'\,\dfrac{d\psi'}{dt}.$$

Writing Ω for $\psi - \alpha$, we obtain

$$\cos\Omega\sin\theta\,\dfrac{d\psi}{dt} = (\sin I' \cos\psi'$$

$$+ \sin i \sin I' \sin\psi' \cot\theta \sin\Omega)\,\dfrac{d\psi'}{dt}$$

$$= (\sin I' \cos\psi' + \sin i \cos\theta \sin^2\Omega)\,\dfrac{d\psi'}{dt}\ \ldots\ldots \text{ by (1)},$$

$$= \left(\dfrac{\cos i \cos I' - \cos\theta}{\sin i}\right.$$

$$\left. + \sin i \cos\theta - \sin i \cos\theta \cos^2\Omega\right)\dfrac{d\psi'}{dt}\ \ldots\ldots \text{ by (2)},$$

$$= \left(\dfrac{\cos i \cos I' - \cos\theta \cos^2 i}{\sin i} - \sin i \cos\theta \cos^2\Omega\right)\dfrac{d\psi'}{dt}$$

$$= (\cos i \sin\theta \cos\Omega - \sin i \cos\theta \cos^2\Omega)\dfrac{d\psi'}{dt},\ \ldots\ldots \text{ by (3)};$$

therefore $\quad \dfrac{d\psi}{dt} = (\cos i - \sin i \cot\theta \cos\Omega)\dfrac{d\psi'}{dt}.$

Also $\quad \dfrac{d\theta}{dt} = -\dfrac{\sin i \sin I' \sin\psi'}{\sin\theta}\dfrac{d\psi'}{dt}$

$$= -\sin i \sin\Omega\,\dfrac{d\psi'}{dt}\ \ldots\ldots\ldots\ldots \text{ by (1)}.$$

We must now substitute the value of $\dfrac{d\psi'}{dt}$: thus

$$\frac{d\psi}{dt} = \frac{3n''^2}{2n(1+\lambda)} \cdot \frac{C-A}{C} \cdot \cos I' (\cos i - \sin i \cot \theta \cos \Omega),$$

$$\frac{d\theta}{dt} = -\frac{3n''^2}{2n(1+\lambda)} \cdot \frac{C-A}{C} \cdot \cos I' \sin i \sin \Omega.$$

It remains to eliminate I'. Now

$$\cos I' (\cos i - \sin i \cot \theta \cos \Omega)$$

$$= (\cos \theta \cos i + \sin \theta \sin i \cos \Omega)(\cos i - \sin i \cot \theta \cos \Omega)$$

$$= \cos \theta \cos^2 i - \frac{1}{2} \frac{\cos 2\theta}{\sin \theta} \sin 2i \cos \Omega - \cos \theta \sin^2 i \cos^2 \Omega$$

$$= \cos \theta \left(\cos^2 i - \frac{1}{2} \sin^2 i \right) - \frac{1}{2} \frac{\cos 2\theta}{\sin \theta} \sin 2i \cos \Omega$$

$$- \frac{1}{2} \cos \theta \sin^2 i \cos 2\Omega.$$

Also $\cos I' \sin i \sin \Omega$

$$= (\cos \theta \cos i + \sin \theta \sin i \cos \Omega) \sin i \sin \Omega$$

$$= \frac{1}{2} \cos \theta \sin 2i \sin \Omega + \frac{1}{2} \sin \theta \sin^2 i \sin 2\Omega.$$

Hence $\dfrac{d\psi}{dt} = \dfrac{3n''^2}{2n(1+\lambda)} \cdot \dfrac{C-A}{C} \cdot \cos \theta (\cos^2 i$

$$- \frac{1}{2} \sin^2 i - \cot 2\theta \sin 2i \cos \Omega - \frac{1}{2} \sin^2 i \cos 2\Omega),$$

$$\frac{d\theta}{dt} = -\frac{3n''^2}{4n(1+\lambda)} \cdot \frac{C-A}{C} \cdot (\cos \theta \sin 2i \sin \Omega$$

$$+ \sin \theta \sin^2 i \sin 2\Omega).$$

Let ν be the mean retrograde velocity of the Moon's nodes, Ω_0 the value of Ω at the epoch, then we may write $\Omega = \nu t + \Omega_0$; also for θ we may write its mean value I; and since i is a very small angle we may neglect $\sin^2 i$ when multiplied by a periodical term. Thus

C. 4

$$\frac{d\psi}{dt} = \frac{3n''^2}{2n(1+\lambda)} \cdot \frac{C-A}{C} \cdot \cos I \left\{ \cos^2 i - \frac{1}{2} \sin^2 i \right.$$
$$\left. - \cot 2I \sin 2i \cos (\nu t + \Omega_0) \right\},$$

$$\frac{d\theta}{dt} = -\frac{3n''^2}{4n(1+\lambda)} \cdot \frac{C-A}{C} \cdot \cos I \sin 2i \sin (\nu t + \Omega_0).$$

The integrals of these equations are

$$\psi = \frac{3n''^2}{2n(1+\lambda)} \cdot \frac{C-A}{C} \cdot \cos I \left\{ \left(\cos^2 i - \frac{1}{2} \sin^2 i \right) t \right.$$
$$\left. - \frac{1}{\nu} \cot 2I \sin 2i \sin (\nu t + \Omega) \right\} + \text{const.},$$

$$\theta = I + \frac{3n''^2}{4n(1+\lambda)} \cdot \frac{C-A}{C\nu} \cos I \sin 2i \cos (\nu t + \Omega_0).$$

38. These equations determine the motion of the Earth's axis due to the attraction of the Moon. They are similar to the equations expressing the Sun's action, and the remarks made in Art. 35 might, *mutatis mutandis*, be repeated here. The Lunar Precession is

$$\frac{3n''^2}{2n(1+\lambda)} \cdot \frac{C-A}{C} \cdot \cos I \left(\cos^2 i - \frac{1}{2} \sin^2 i \right) t.$$

If we add to this the Solar Precession (Art. 32), we find for the whole permanent effect of the Sun and Moon upon the equinoxes

$$\frac{3n}{2} \cdot \frac{C-A}{C} \cdot \cos I \left\{ \frac{n''^2}{n^2(1+\lambda)} \left(\cos^2 i - \frac{1}{2} \sin^2 i \right) + \frac{n'^2}{n^2} \right\} t.$$

This is called the *Luni-solar Precession*, to distinguish it from the Precession due to the secular motion of the ecliptic in consequence of the attractions of the Planets, and which is called *Planetary Precession*.

Adding together the effects of Lunar and Solar Nutation, we find for the whole Nutation in Longitude

$$-\frac{3}{2} \cdot \frac{C-A}{C} \cdot \cos I \left\{ \frac{n''^2}{n\nu\,(1+\lambda)} \cot 2I \sin 2i \sin \Omega \right.$$

$$\left. + \frac{n'}{2n} \sin 2\odot \right\},$$

where Ω and $\odot$ denote respectively the mean longitude of the ascending node of the Moon's orbit and of the Sun.

Similarly, the whole Nutation in Latitude is

$$\frac{3}{4} \cdot \frac{C-A}{C} \cdot \cos I \left\{ \frac{n''^2}{n\nu\,(1+\lambda)} \sin 2i \cos \Omega + \frac{n'}{n} \cos 2\odot \right\}.$$

In each of these expressions the first term, due to the action of the Moon, is the most important, since n'' is larger than both n' and ν.

39. If we consider only the motion of the Earth's axis due to Precession, it appears from the preceding formulæ that it maintains a constant inclination to the pole of the ecliptic, and describes a right circular cone about it with uniform velocity. An axis possessing this motion exactly we shall term the *mean axis* of the Earth.

The motion of the axis due to Lunar Nutation can now be exhibited as follows:—

Let $x = -\dfrac{3}{4} \cdot \dfrac{n''^2}{n\nu\,(1+\lambda)} \cdot \dfrac{C-A}{C} \cdot \cos 2I \sin 2i \sin \Omega,$

$y = \dfrac{3}{4} \cdot \dfrac{n''^2}{n\nu\,(1+\lambda)} \cdot \dfrac{C-A}{C} \cdot \cos I \sin 2i \cos \Omega:$

then, taking unity as the radius of the celestial sphere, x and y are the small *linear* spaces traversed by the intersection of the Earth's axis with the circumference in two directions at right angles. Eliminating Ω from these equations, we have

$$\frac{x^2}{\cos^2 2I} + \frac{y^2}{\cos^2 I} = \left(\frac{3}{4} \frac{n''^2}{n\nu\,(1+\lambda)} \cdot \frac{C-A}{C} \cdot \sin 2i \right)^2.$$

Hence, in consequence of Lunar Nutation, the extremity of the axis may be considered to move in an ellipse whose semi-axes are in the ratio of $\cos 2I$ to $\cos I$. The centre of this ellipse is at the point of intersection of this mean axis with the celestial sphere, and its plane a tangent to the sphere at that point. By supposing the mean axis to describe the cone uniformly, while the true axis describes this ellipse about it, the real motion will be represented. This conception is due to Bradley, who arrived at it by observation.

40. The annual value of the Luni-solar precession (see Art. 38)

$$= \frac{3n}{2n'} \cdot \frac{C-A}{C} \cos I \left\{ \frac{n''^2}{n^2(1+\lambda)} \left(\cos^2 i - \frac{1}{2} \sin^2 i \right) + \frac{n'^2}{n^2} \right\} n'.$$

Observation gives about $50''.1$ as the numerical value of this expression. Hence, by substituting the known values of n, n', ν, I, i, we have a relation between $\dfrac{C-A}{C}$ and λ, by means of which either may be determined when the other is known.

41. We have taken no account in our calculations of the Precession caused by the attractions of the planets on the Earth, since it is too trifling to be appreciable.

THE END.

CAMBRIDGE: PRINTED AT THE UNIVERSITY PRESS.

MATHEMATICAL WORKS

PUBLISHED BY

MACMILLAN AND CO.

An Elementary Treatise on the Planetary Theory. With
a Collection of Problems. By C. H. H. CHEYNE, B.A. Crown 8vo.
6s. 6d.

A Treatise on Astronomy, for the Use of Colleges and
Schools. By HUGH GODFRAY, M.A. 8vo. 12s. 6d.

An Elementary Treatise on the Lunar Theory. With a
brief Sketch of the Problem up to the time of Newton. By HUGH
GODFRAY, M.A. Second Edition revised. Crown 8vo. 5s. 6d.

The Singular Properties of the Ellipsoid and Associated
Surfaces of the nth Degree. By the Rev. G. F. CHILDE, M.A. 8vo.
10s. 6d.

Works by G. B. AIRY, M.A. LL.D. D.C.L.
Astronomer Royal, &c.

Treatise on the Algebraical and Numerical Theory of Errors
of Observations and the Combination of Observations. Crown 8vo.
6s. 6d.

An Elementary Treatise on Partial Differential Equations.
With Stereoscopic Cards of Diagrams. Crown 8vo. 5s. 6d.

On the Undulatory Theory of Optics. Designed for the
use of Students in the University. Crown 8vo. 6s. 6d.

MACMILLAN AND CO. LONDON.

Dynamics of a Particle. With Examples. By PROFESSOR TAIT and MR STEELE. New Edition. Crown 8vo. 10s. 6d.

Dynamics of a System of Rigid Bodies. With Examples. By EDWARD JOHN ROUTH. Crown 8vo. 10s. 6d.

Elementary Hydrostatics. By J. B. PHEAR, M.A. Fourth Edition. Crown 8vo. 5s. 6d.

The Elements of Molecular Mechanics. By JOSEPH BAYMA, S.J., Professor of Philosophy, Stonyhurst College. Demy 8vo. 10s. 6d.

Works by the late GEORGE BOOLE, F.R.S. *Professor of Mathematics in the Queen's University, Ireland, &c.*

A Treatise on Differential Equations. New Edition. Edited by I. TODHUNTER, M.A. F.R.S. 8vo. cloth. 14s.

Treatise on Differential Equations. Supplementary Volume. Crown 8vo. 8s. 6d.

A Treatise on the Calculus of Finite Differences. Crown 8vo. 10s. 6d.

MACMILLAN AND CO. LONDON.

The First Three Sections of Newton's Principia. With Notes and Problems. By PERCIVAL FROST, M.A. Second Edition. 8vo. 10s. 6d.

A Treatise on Solid Geometry. By the Rev. PERCIVAL FROST, M.A., and the Rev. J. WOLSTENHOLME, M.A. 8vo. 18s.

A Collection of Mathematical Problems and Examples. By H. A. MORGAN, M.A. Crown 8vo. 6s. 6d.

Conic Sections and Algebraic Geometry. With Easy Examples, progressively arranged. By G. H. PUCKLE. New and Revised Edition. Crown 8vo. 7s. 6d. *[In the Press.*

An Elementary Treatise on Trilinear Co-ordinates, the Method of Reciprocal Polars, and the Theory of Projections. By N. M. FERRERS. Second Edition. Crown 8vo. 6s. 6d.

A Treatise on Optics. By the Rev. S. PARKINSON, B.D. Crown 8vo. New Edition. 10s. 6d.

A Collection of Elementary Test-Questions in Pure and Mixed Mathematics, with Answers and Appendices on Synthetic Division, and on the Solution of Numerical Equations by Horner's Method. By JAMES R. CHRISTIE, F.R.S. Crown 8vo. 8s. 6d.

MACMILLAN AND CO. LONDON.

Works by ISAAC TODHUNTER, M.A. F.R.S.

A Treatise on the Differential Calculus. With Examples.
Fourth Edition. Crown 8vo. 10s. 6d.

A Treatise on the Integral Calculus. Second Edition, with
Examples. Crown 8vo. 10s. 6d.

A Treatise on Analytical Statics. With Examples. Third
Edition. Crown 8vo. 10s. 6d.

A Treatise on Conic Sections. With Examples. Fourth
and Cheaper Edition. Crown 8vo. 7s. 6d.

Examples of Analytical Geometry of Three Dimensions.
Second Edition. Crown 8vo. 4s.

A Treatise on the Theory of Equations. Second Edition.
Crown 8vo. 7s. 6d.

Mathematical Theory of Probability. 8vo. cloth. 18s.

MACMILLAN AND CO. LONDON.

MACMILLAN AND CO.'S

LIST OF PUBLICATIONS.

A BOOK OF THOUGHTS. By H. A. 18mo. cloth extra, gilt, 3*s.* 6*d.*

ACROSS THE CARPA-THIANS. In 1858-60. With a Map. Crown 8vo. 7*s.* 6*d.*

ÆSCHYLI EUMENIDES. The Greek Text with English Notes, and an Introduction. By BERNARD DRAKE, M.A. 8vo. 7*s.* 6*d.*

AGNES HOPETOUN. 16mo. cloth. *See* OLIPHANT.

AIRY.—TREATISE on the ALGEBRAICAL and NUME-RICAL THEORY of ERRORS of OBSERVATIONS, and the Combination of Observations. By G. B. AIRY, M.A. Crown 8vo. 6*s.* 6*d.*

AIRY. — POPULAR AS-TRONOMY: A Series of Lectures delivered at Ipswich. By GEORGE BIDDELL AIRY, Astronomer Royal. 18mo. cloth. Uniform with Macmillan's School Class Books. [*In the Press.*

ALGEBRAICAL EXER-CISES. Progressively arranged by Rev. C. A. JONES, M.A. and C. H. CHEYNE, M.A. Mathematical Masters in Westminster School. Pott 8vo. cl., price 2*s.* 6*d.*

ALICE'S ADVENTURES in WONDERLAND. With Forty-two Illustrations by TENNIEL. Crown 8vo. cloth, 7*s.* 6*d.*

ALLINGHAM.—
LAURENCE BLOOMFIELD in IRELAND. A Modern Poem. By WILLIAM ALLINGHAM. Fcap. 8vo. 7*s.*

ANDERSON. — SEVEN MONTHS' RESIDENCE in RUSSIAN POLAND in 1863. By the Rev. FORTESCUE L. M. ANDERSON. Cr. 8vo. 6*s.*

ANOTHER "STORY of the GUNS;" or Sir Emerson Tennent and the Whitworth Gun. By the "FRASER REVIEWER." Extra fcap. 8vo. 2*s.*

ANSTED.—THE GREAT STONE BOOK of NATURE. By DAVID THOS. ANSTED, M.A. F.R.S. F.G.S. Fcap. 8vo. 5*s.*

ANSTIE. — STIMULANTS and NARCOTICS, their MU-TUAL RELATIONS. With Special Researches on the Action of Alcohol, Æther, and Chloroform, on the Vital Organism. By FRANCIS E. ANSTIE, M.D. M.R.C.P. 8vo. 14*s.*

▲

ARISTOTLE ON THE VITAL PRINCIPLE. Translated, with Notes, by CHARLES COLLIER, M.D. F.R.S. Cr. 8vo. 8s. 6d.

ARNOLD. — A FRENCH ETON ; or, Middle Class Education and the State. By MATTHEW ARNOLD. Fcap. 8vo. 2s. 6d.

ARNOLD. — ESSAYS ON CRITICISM. By MATTHEW ARNOLD, Professor of Poetry in the University of Oxford. Extra fcp. 8vo. cloth, 6s.

ARTIST and CRAFTSMAN. A Novel. Crown 8vo. 6s.

BARWELL.—GUIDE in the SICK ROOM. By RICHARD BARWELL, F.R.C.S. Extra fcap. 8vo. 3s. 6d.

BEASLEY.—An ELEMENTARY TREATISE on PLANE TRIGONOMETRY. With a Numerous Collection of Examples. By R. D. BEASLEY, M.A. Crown 8vo. 3s. 6d.

BELL.—ROMANCES AND MINOR POEMS. By HENRY GLASSFORD BELL. Fcap. 8vo.

BIRKS.—The DIFFICULTIES of BELIEF in connexion with the CREATION and the FALL. By THOS. RAWSON BIRKS, M.A. Cr. 8vo. 4s. 6d.

BIRKS.—On MATTER and ETHER ; or the Secret Laws of Physical Change. By THOS. RAWSON BIRKS, M.A. Cr. 8vo. 5s. 6d.

BLAKE.—THE LIFE OF WILLIAM BLAKE, the Artist. By ALEXR. GILCHRIST. With numerous Illustrations from Blake's Designs and Fac-similes of his Studies of the "Book of Job." 2 vols. Medium 8vo. 32s.

BLANCHE LISLE, AND OTHER POEMS. By CECIL HOME. Fcap. 8vo. 4s. 6d.

BOOLE.—A TREATISE on DIFFERENTIAL EQUATIONS. By GEO. BOOLE, D.C.L. New Edition. Edited by I. TODHUNTER, M.A., F.R.S. 8vo. cloth, 14s.

BOOLE.—A TREATISE on the CALCULUS of FINITE DIFFERENCES. By GEO. BOOLE, D.C.L. Crown 8vo. 10s. 6d.

BRADSHAW. — AN ATTEMPT TO ASCERTAIN THE STATE of CHAUCER'S WORKS, as they were Left at his Death. With some Notice of their Subsequent History. By HENRY BRADSHAW, o King's College, and the University Library, Cambridge.
[In the Press

BRIEF BIOGRAPHICAL DICTIONARY. See HOLE.

BRIMLEY.—ESSAYS, by the late GEORGE BRIMLEY, M.A. Edited by W. G. CLARK, M.A. With Portrait. Second Edition, Fcap. 8vo. 5s.

BROCK.—DAILY READINGS on the PASSION of OUR LORD. By Mrs. H. F. BROCK Fcap. 8vo. 4s.

BROOK SMITH.—ARITH-
METIC in THEORY and
PRACTICE. For Advanced
Pupils. Part First. By J. BROOK
SMITH, M.A. Crown 8vo.
3s. 6d.

BROTHER FABIAN'S
MANUSCRIPT. See EVANS.

BRYCE. — THE HOLY
ROMAN EMPIRE. By JAMES
BRYCE, B.A. Fellow of Oriel
College, Oxford. A New Edi-
tion, revised. Crown 8vo.
[In the Press.

BULLOCK—POLISH EX-
PERIENCES during the IN-
SURRECTION of 1863-4. By
W. H. BULLOCK. Crown 8vo.
with Map, 8s. 6d.

BURGON.—A TREATISE
on the PASTORAL OFFICE.
Addressed chiefly to Candidates
for Holy Orders, or to those who
have recently undertaken the cure
of souls. By the Rev. JOHN W.
BURGON, M.A. 8vo. 12s.

BUTLER (ARCHER). —
WORKS by the Rev. WILLIAM
ARCHER BUTLER, M.A. late
Professor of Moral Philosophy in
the University of Dublin :—

1. SERMONS, DOCTRINAL
and PRACTICAL. Edited, with
a Memoir of the Author's Life, by
THOMAS WOODWARD, M.A.
With Portrait. *Sixth Edition.*
8vo. 12s.

2. A SECOND SERIES OF
SERMONS. Edited by J. A.
JEREMIE, D.D. *Third Edition.*
8vo. 10s. 6d.

3. HISTORY OF ANCIENT
PHILOSOPHY. Edited by
WM. H. THOMPSON, M.A.
2 vols. 8vo. 1l. 5s.

4. LETTERS ON ROMANISM,
in REPLY to MR. NEWMAN'S
ESSAY on DEVELOPMENT.
Edited by the Very Rev. T.
WOODWARD. *Second Edition,*
revised by Archdeacon HARD-
WICK. 8vo. 10s. 6d.

BUTLER (MONTAGU).—
SERMONS PREACHED in
the CHAPEL of HARROW
SCHOOL. By H. MONTAGU
BUTLER, Head Master. Crown
8vo. 7s. 6d.

BUTLER. — FAMILY
PRAYERS. By the Rev. GEO.
BUTLER. Cr. 8vo. 5s.

BUTLER. — SERMONS
PREACHED in CHELTEN-
HAM COLLEGE. CHAPEL.
By the Rev. GEO. BUTLER.
Crown 8vo. 7s. 6d.

CAIRNES.—THE SLAVE
POWER; its Character, Career,
and Probable Designs. Being an
Attempt to Explain the Real
Issues Involved in the American
Contest. By J. E. CAIRNES,
M.A. *Second Edition.* 8vo.
10s. 6d.

CALDERWOOD.—PHILO-
SOPHY of the INFINITE. A
Treatise on Man's Knowledge of
the Infinite Being, in answer to
Sir W. Hamilton and Dr. Mansel.
By the Rev. HENRY CALDER-
WOOD, M.A. *Second Edition.*
8vo. 14s.

ST. PAUL'S EPISTLE TO THE ROMANS. Newly Translated and Explained, from a Missionary point of View. Crown 8vo. 7s. 6d.

LETTER TO HIS GRACE THE ARCHBISHOP OF CANTERBURY, upon the Question of Polygamy, as found already existing in Converts from Heathenism. Second Edition. Crown 8vo. 1s. 6d.

COOKERY FOR ENGLISH HOUSEHOLDS. By a FRENCH LADY. Extra fcap. 8vo. 5s.

COOPER.—ATHENAE CANTABRIGIENSES. By CHARLES HENRY COOPER, F. S. A. and THOMPSON COOPER, F.S.A. Vol. I. 8vo. 1500—85, 18s. Vol. II. 1586—1609, 18s.

COTTON.—SERMONS and ADDRESSES delivered in Marlborough College during Six Years, by GEORGE EDWARD LYNCH COTTON, D.D. Lord Bishop of Calcutta. Crown 8vo. 10s. 6d.

COTTON.—A CHARGE to the CLERGY of the DIOCESE and PROVINCEofCALCUTTA at the Second Diocesan and First Metropolitan Visitation. By GEORGE EDWARD LYNCH COTTON, D.D. 8vo. 3s. 6d.

COTTON.—SERMONS, chiefly connected with Public Events of 1854. Fcap. 8vo. 3s.

COTTON.—EXPOSITORY SERMONS on the EPISTLES for the Sundays of the Christian Year. By GEORGE EDWARD LYNCH COTTON, D.D. Two Vols. crown 8vo. 15s.

CRAIK.—MY FIRST JOURNAL. A book for the Young. By GEORGIANA M. CRAIK, Author of "Riverston," "Lost and Won," &c. Royal 16mo. cloth, gilt leaves, 3s. 6d.

CROCKER.—A NEW PROPOSAL for a GEOGRAPHICAL SYSTEM of MEASURES and WEIGHTS conveniently Introducible, generally by retaining familiar notions by familiar names. To which are added remarks on systems of Coinage. By JAMES CROCKER, M.A. 8vo. 8s. 6d.

DANTE. — DANTE'S COMEDY, *The Hell.* Translated by W. M. ROSETTI. Fcap. 8vo. cloth, 5s.

DAVIES.—ST. PAUL AND MODERN THOUGHT: Remarks on some of the Views advanced in Professor Jowett's Commentary on St. Paul. By Rev. J. LL. DAVIES, M.A. 8vo. 2s. 6d.

DAVIES.—SERMONS ON THE MANIFESTATION OF THE SON OF GOD. With a Preface addressed to Laymen on the present position of the Clergy of the Church of England; and an Appendix on the Testimony of Scripture and the Church as to the possibility of Pardon in the Future State. By the Rev. J. LL. DAVIES, M.A. Fcap. 8vo. 6s. 6d.

DAVIES.—THE WORK OF CHRIST; OR THE WORLD RECONCILED TO GOD. With a Preface on the Atonement Controversy. By the Rev. J. LL. DAVIES, M.A. Fcap. 8vo. 6s.

DAVIES.—BAPTISM, CONFIRMATION, AND THE LORD'S SUPPER, as interpreted by their outward signs. Three Expository Addresses for Parochial Use. By the Rev. J. LL. DAVIES, M.A. Limp cloth, 1s. 6d.

DAYS OF OLD : STORIES FROM OLD ENGLISH HISTORY. By the Author of "Ruth and her Friends." *New Edition.* 18mo. cloth, gilt leaves, 3s. 6d.

DEMOSTHENES DE CORONA. The Greek Text with English Notes. By B. DRAKE, M.A. *Second Edition,* to which is prefixed AESCHINES AGAINST CTESIPHON, with English Notes. Fcap. 8vo. 5s.

DE TEISSIER.—VILLAGE SERMONS, by G. F. DE TEISSIER, B.D. Crown 8vo. 9s. SECOND SERIES. Crown 8vo. cloth. 8s. 6d.

DE VERE.—THE INFANT BRIDAL, AND OTHER POEMS. By AUBREY DE VERE. Fcap. 8vo. 7s. 6d.

DICEY. — SIX MONTHS IN THE FEDERAL STATES. By EDWARD DICEY. 2 Vols. crown 8vo. 12s.

DICEY.—ROME IN 1860. By EDWARD DICEY. Crown 8vo. 6s. 6d.

DONALDSON.—A CRITICAL HISTORY OF CHRISTIAN LITERATURE AND DOCTRINE, from the Death of the Apostles to the Nicene Council. By JAMES DONALDSON, M.A. Vol. I.—THE APOSTOLIC FATHERS. 8vo. cloth. 10s. 6d.
Vols. II. and III. just ready.

DREW. — A GEOMETRICAL TREATISE ON CONIC SECTIONS. By W. H. DREW, M.A. *Third Edition.* Crown 8vo. 4s. 6d.

DREW.—SOLUTIONS TO PROBLEMS CONTAINED IN MR. DREW'S TREATISE ON CONIC SECTIONS. Crown 8vo. 4s. 6d.

EARLY EGYPTIAN HISTORY FOR THE YOUNG. With Descriptions of the Tombs and Monuments. *New Edition,* with Frontispiece. Fcap. 8vo. 5s.

EASTWOOD.—The BIBLE WORD BOOK. A Glossary of Old English Bible Words. By J. EASTWOOD, M.A. of St. John's College, and W. ALDIS WRIGHT, M.A. Trinity College, Cambridge. 18mo. Uniform with Macmillan's School Class Books.

ECCE HOMO. A Survey of the Life and Work of Jesus Christ. 8vo. 10s. 6d.

ECHOES OF MANY VOICES FROM MANY LANDS. 18mo. cloth extra, gilt, 3s. 6d.

ENGLISH IDYLLS. By JANE ELLICE. Fcap. 8vo. cloth. 6s.

EVANS.—BROTHER FABIAN'S MANUSCRIPT; And other Poems. By SEBASTIAN EVANS. Fcap. 8vo. cloth, price 6s.

FAWCETT. — THE ECONOMIC POSITION OF THE BRITISH LABOURER. By HENRY FAWCETT, M.P. Extra fcap. 8vo. cloth, 5s.

FAWCETT.—MANUAL of POLITICAL ECONOMY. By HENRY FAWCETT, M.P. *Second Edition.* Crown 8vo. 12*s.*

FERRERS.—A TREATISE ON TRILINEAR CO-ORDI-NATES, the Method of Reciprocal Polars, and the Theory of Projections. By the Rev. N. M. FERRERS, M.A. Crown 8vo. 6*s.* 6*d.*

FISHER. — CONSIDERA-TIONS ON THE ORIGIN OF THE AMERICAN WAR. By HERBERT FISHER. Fcp. 8vo. 2*s.* 6*d.*

FLETCHER. — THOUGHTS FROM A GIRL'S LIFE. By LUCY FLETCHER. *Second Edition.* Fcap. 8vo. 4*s.* 6*d.*

FORBES.—LIFE OF EDWARD FORBES, F.R.S. By GEORGE WILSON, M.D. F.R.S.E. and ARCHIBALD GEIKIE, F.G.S. 8vo. with Portrait, 14*s.*

FREEMAN.—HISTORY of FEDERAL GOVERNMENT, from the Foundation of the Achaian League to the Disruption of the United States. By EDWARD A. FREEMAN, M.A. Vol. I. General Introduction.—History of the Greek Federations. 8vo. 21*s.*

FROST.—THE FIRST THREE SECTIONS of NEW-TON'S PRINCIPIA. With Notes and Problems in illustration of the subject. By PERCIVAL FROST, M.A. *Second Edition.* 8vo. 10*s.* 6*d.*

FROST AND WOLSTEN-HOLME.—A TREATISE ON SOLID GEOMETRY. By the Rev. PERCIVAL FROST, M.A. and the Rev. J. WOLSTEN-HOLME, M.A. 8vo. 18*s.*

FURNIVALL.–LE MORTE ARTHUR. Edited from the Harleian M.S. 2252, in the British Museum. By F. J. FURNI-VALL, M.A. With Essay by the late HERBERT COLE-RIDGE. Fcap. 8vo. cloth, 7*s.* 6*d.*

GALTON.—METEORO-GRAPHICA, or Methods of Mapping the Weather. Illustrated by upwards of 600 Printed Lithographed Diagrams. By FRAN-CIS GALTON, F.R.S. 4to. 9*s.*

GARIBALDI at CAPRERA. By COLONEL VECCHJ. With Preface by Mrs. GASKILL. Fcap. 8vo. 1*s.* 6*d.*

GEIKIE.—STORY OF A BOULDER; or, Gleanings by a Field Geologist. By ARCHI-BALD GEIKIE. Illustrated with Woodcuts. Crown 8vo. 5*s.*

GEIKIE'S SCENERY OF SCOTLAND, with Illustrations and a new Geological Map. Cr. 8vo. cloth, 10*s.* 6*d.*

GIFFORD.—THE GLORY OF GOD IN MAN. By E. H. GIFFORD, D.D. Fcap. 8vo. cloth. 3*s.* 6*d.*

GOLDEN TREASURY SERIES. Uniformly printed in 18mo. with Vignette Titles by J. NOEL PATON, T. WOOLNER, W. HOLMAN HUNT, J. E. MIL-LAIS, &c. Bound in extra cloth, 4*s.* 6*d.* ; morocco plain, 7*s.* 6*d.* ; morocco extra, 10*s.* 6*d.* each Volume.

THE GOLDEN TREASURY OF THE BEST SONGS AND LYRICAL POEMS IN THE ENGLISH LANGUAGE. Selected and arranged, with Notes, by FRANCIS TURNER PALGRAVE.

THE CHILDREN'S GARLAND FROM THE BEST POETS. Selected and arranged by COVENTRY PATMORE.

THE BOOK OF PRAISE. From the best English Hymn Writers. Selected and arranged by ROUNDELL PALMER. A New and Enlarged Edition.

THE FAIRY BOOK : The Best Popular Fairy Stories. Selected and Rendered Anew by the Author of "John Halifax."

THE BALLAD BOOK. A Selection of the Choicest British Ballads. Edited by WILLIAM ALLINGHAM.

THE JEST BOOK. The Choicest Anecdotes and Sayings. Selected and arranged by MARK LEMON.

BACON'S ESSAYS AND COLOURS OF GOOD AND EVIL. With Notes and Glossarial Index, by W. ALDIS WRIGHT, M.A. Large paper copies, crown 8vo. 7s. 6d.; or bound in half morocco, 10s. 6d.

The PILGRIM'S PROGRESS from this World to that which is to Come. By JOHN BUNYAN.

₊ Large paper Copies, crown 8vo. cloth, 7s. 6d.; or bound in half morocco, 10s. 6d.

THE SUNDAY BOOK OF POETRY FOR THE YOUNG. Selected and arranged by C. F. ALEXANDER.

A BOOK OF GOLDEN DEEDS OF ALL TIMES AND ALL COUNTRIES. Gathered and Narrated anew by the Author of "The Heir of Redclyffe."

THE POETICAL WORKS OF ROBERT BURNS. Edited, with Biographical Memoir, by ALEXANDER SMITH. 2 vols.

THE ADVENTURES OF ROBINSON CRUSOE. Edited from the Original Editions by J. W. CLARK, M.A. with a Vignette Title by J. E. MILLAIS.

THE REPUBLIC OF PLATO. Translated into English with Notes by J. Ll. DAVIES, M.A. and D. J. VAUGHAN, M.A. New Edition, with Vignette Portraits of Plato and Socrates engraved by JEENS from an Antique Gem.

GORDON. — LETTERS from EGYPT, 1863—5. By LADY DUFF GORDON. *Third Edition.* Cr. 8vo. cloth, 8s. 6d.

GORST.—THE MAORI KING; or, the Story of our Quarrel with the Natives of New Zealand. By J. E. GORST, M.A. With a Portrait of William Thompson, and a Map of the Seat of War. Crown 8vo. 10s. 6d.

GREEN. — SPIRITUAL PHILOSOPHY, Founded on the Teaching of the late SAMUEL TAYLOR COLERIDGE. By the late JOSEPH HENRY GREEN, F.R.S. D.C.L. Edited, with a Memoir of the Author's Life, by JOHN SIMON, F.R.S. Two vols. 8vo. cloth, price, 25s.

GROVES.—A COMMENTARY ON THE BOOK OF GENESIS. For the Use of Students and Readers of the English Version of the Bible. By the Rev. H. C. GROVES, M.A. Crown 8vo. 9s.

GUIDE TO THE UNPRO-TECTED in Every Day Matters relating to Property and Income. By a BANKER'S DAUGHTER. *Second Edition.* Extra fcap. 8vo. 3*s.* 6*d.*

HAMERTON.—A PAINT-ER'S CAMP IN THE HIGH-LANDS; and Thoughts about Art. By P. G. HAMERTON. 2 vols. crown 8vo. 21*s.*

HAMILTON. — THE RE-SOURCES OF A NATION. A Series of Essays. By ROW-LAND HAMILTON. 8vo. 10*s.* 6*d.*

HAMILTON.—On TRUTH and ERROR: Thoughts on the Principles of Truth, and the Causes and Effects of Error. By JOHN HAMILTON. Crown 8vo. 5*s.*

HARDWICK.—CHRIST AND OTHER MASTERS. A Historical Inquiry into some of the Chief Parallelisms and Contrasts between Christianity and the Religious Systems of the Ancient World. By the Ven. ARCH-DEACON HARDWICK. *New Edition*, revised, and a Prefatory Memoir by the Rev. FRANCIS PROCTER. Two vols. crown 8vo. 15*s.*

HARDWICK. — A HIS-TORY OF THE CHRISTIAN CHURCH, during the Middle Ages and the Reformation. (A.D. 590—1600.) By ARCHDEACON HARDWICK. Two vols. crown 8vo. 21*s.*

Vol. I. *Second Edition.* Edited by FRANCIS PROCTER, M.A. History from Gregory the Great to the Excommunication of Luther. With Maps.

Vol. II. *Second Edition.* History of the Reformation of the Church.

Each volume may be had separately. Price 10*s.* 6*d.*

HARDWICK. — TWENTY SERMONS FOR TOWN CON-GREGATIONS. Crown 8vo. 6*s.* 6*d.*

HARE.—WORKS BY JULIUS CHARLES HARE, M.A. Sometime Archdeacon of Lewes, and Chaplain in Ordinary to the Queen.

CHARGES DELIVERED during the Years 1840 to 1854. With Notes on the Principal Events affecting the Church during that period. With an Introduction, explanatory of his position in the Church with reference to the parties which divide it. 3 vols. 8vo. 1*l.* 11*s.* 6*d.*

MISCELLANEOUS PAM-PHLETS on some of the Leading Questions agitated in the Church during the Years 1845—51. 8vo. 12*s.*

THE VICTORY OF FAITH. 5*s.*

THE MISSION OF THE COM-FORTER. *Third Edition.* With Notes, 12*s.*

VINDICATION OF LUTHER. *Second Edition.* 8vo. 7*s.*

PARISH SERMONS. Second Series. 8vo. cloth, 12*s.*

SERMONS PREACHED ON PARTICULAR OCCASIONS. 8vo. cloth, 12*s.*

PORTIONS OF THE PSALMS IN ENGLISH VERSE. Selected for Public Worship. 18mo. cloth, 2*s.* 6*d.*

*** The two following Books are included in the Three Volumes of Charges, but may be had separately.

THE CONTEST WITH ROME. *Second Edition.* 8vo. cl. 10s. 6d.

CHARGES DELIVERED in the Years 1843, 1845, 1846. With an Introduction. 6s. 6d.

HEARN. — PLUTOLOGY; or, the Theory of the Efforts to Satisfy Human Wants. By W. E. HEARN, LL.D. 8vo. 14s.

HEBERT. — CLERICAL SUBSCRIPTION, an Inquiry into the Real Position of the Church and the Clergy in reference to — I. The Articles; II. The Liturgy; III. The Canons and Statutes. By the Rev. CHARLES HEBERT, M.A. F.R.S.L. Cr. 8vo. 7s. 6d.

HEMMING. — AN ELEMENTARY TREATISE ON THE DIFFERENTIAL AND INTEGRAL CALCULUS. By G. W. HEMMING, M.A. *Second Edition.* 8vo. 9s.

HERVEY.—THE GENEALOGIES OF OUR LORD AND SAVIOUR JESUS CHRIST, as contained in the Gospels of St. Matthew and St. Luke, reconciled with each other, and shown to be in harmony with the true Chronology of the Times. By Lord ARTHUR HERVEY, M.A. 8vo. 10s. 6d.

HERVEY. — THE AARBERGS. By ROSAMOND HERVEY. 2 vols. crown 8vo. cloth, 21s.

HISTORICUS.—LETTERS ON SOME QUESTIONS OF INTERNATIONAL LAW. Reprinted from the *Times,* with considerable Additions. 8vo. 7s. 6d. Also, ADDITIONAL LETTERS. 8vo. 2s. 6d.

HODGSON.—MYTHOLOGY FOR LATIN VERSIFICATION : a Brief Sketch of the Fables of the Ancients, prepared to be rendered into Latin Verse for Schools. By F. HODGSON, B.D. late Provost of Eton. New Edition, revised by F. C. HODGSON, M.A. 18mo. 3s.

HOLE. — A BRIEF BIOGRAPHICAL DICTIONARY. Compiled and Arranged by CHARLES HOLE, M.A. Trinity College, Cambridge. In Pott 8vo. (same size as the "Golden Treasury Series") neatly and strongly bound, in cloth. *Second Edition.* Price 4s. 6d.

HORNER.—The TUSCAN POET GIUSEPPE GIUSTI AND HIS TIMES. By SUSAN HORNER. Crown 8vo. 7s. 6d.

HOWARD.—THE PENTATEUCH; or, the Five Books of Moses. Translated into English from the Version of the LXX. With Notes on its Omissions and Insertions, and also on the Passages in which it differs from the Authorized Version. By the Hon. HENRY HOWARD, D.D. Crown 8vo. GENESIS, 1 vol. 8s. 6d.; EXODUS AND LEVITICUS, 1 vol. 10s. 6d.; NUMBERS AND DEUTERONOMY, 1 vol. 10s. 6d.

HUMPHRY. — THE HUMAN SKELETON (including the JOINTS). By GEORGE MURRAY HUMPHRY, M.D. F.R.S. With Two Hundred and Sixty Illustrations drawn from Nature. Medium 8vo. 1l. 8s.

HUMPHRY. — THE HUMAN HAND AND THE HUMAN FOOT. With Numerous Illustrations. Fcp. 8vo. 4s. 6d.

HUXLEY. — LESSONS IN ELEMENTARY PHYSIOLO-GY. With numerous Illustrations. By T. H. HUXLEY, F.R.S. Professor of Natural History in the Government School of Mines. Uniform with Macmillan's School Class Books.

HYDE. — HOW TO WIN OUR WORKERS. An Account of the Leeds Sewing School. By Mrs. HYDE. Fcap. 8vo. 1s. 6d.

HYMNI ECCLESIÆ.— Fcap. 8vo. cloth, 7s. 6d.

JAMESON.—LIFE'S WORK, IN PREPARATION AND IN RETROSPECT. Sermons Preached before the University of Cambridge. By the Rev. F. J. JAMESON, M.A. Fcap. 8vo. 1s. 6d.

JAMESON.—BROTHERLY COUNSELS TO STUDENTS. Sermons preached in the Chapel of St. Catharine's College, Cambridge. By F. J. JAMESON, M.A. Fcap. 8vo. 1s. 6d.

JANET'S HOME.—A Novel. *New Edition.* Crown 8vo. 6s.

JEVONS.—THE COAL QUESTION. By W. STANLEY JEVONS, M.A. Fellow of University College, London. 8vo. 10s. 6d.

JONES.—THE CHURCH of ENGLAND and COMMON SENSE. By HARRY JONES, M.A. Fcap. 8vo. cloth, 3s. 6d.

JUVENAL.—JUVENAL, for Schools. With English Notes. By J. E. B. MAYOR, M.A. *New and Cheaper Edition.* Crown 8vo. *Reprinting.*

KEARY. — THE LITTLE WANDERLIN, and other Fairy Tales. By A. and E. KEARY. 18mo cloth, 3s. 6d.

KINGSLEY.—WORKS BY THE REV. CHARLES KINGSLEY, M.A. Rector of Eversley, and Professor of Modern History in the University of Cambridge :—

THE ROMAN and the TEUTON. A Series of Lectures delivered before the University of Cambridge. 8vo. 12s.

TWO YEARS AGO. *Fourth Edition.* Crown 8vo. 6s.

"WESTWARD HO!" *Fifth Edition.* Crown 8vo. 6s.

ALTON LOCKE. *New Edition.* with a New Preface. Crown 8vo. 4s. 6d.

HYPATIA; *Fourth Edition.* Crn. 8vo. 6s.

YEAST. *Fourth Edition.* Fcap. 8vo. 5s.

MISCELLANIES. *Second Edition.* 2 vols. crown 8vo. 12s.

THE SAINT'S TRAGEDY. *Third Edition.* Fcap. 8vo. 5s.

ANDROMEDA, and other Poems. *Third Edition.* Fcap. 8vo. 5s.

THE WATER BABIES, a Fairy Tale for a Land Baby. With Two Illustrations by J. NOEL PATON, R.S.A. *New Edition.* Crown 8vo. 6s.

GLAUCUS : or, the Wonders of the Shore. *New and Illustrated Edition*, containing beautifully Coloured Illustrations. 5s.

THE HEROES ; or, Greek Fairy Tales for my Children. With Eight Illustrations. *New Edition.* 18mo. 3s. 6d.

VILLAGE SERMONS. *Sixth Edition.* Fcap. 8vo. 2*s.* 6*d.*

THE GOSPEL OF THE PENTATEUCH. *Second Edition.* Fcap. 8vo. 4*s.* 6*d.*

GOOD NEWS OF GOD. *Third Edition.* Fcap. 8vo. 6*s.*

SERMONS FOR THE TIMES. *Third Edition.* Fcap. 8vo. 3*s.* 6*d.*

TOWN AND COUNTRY SERMONS. Fcap. 8vo. 6*s.*

SERMONS ON NATIONAL SUBJECTS. First Series. *Second Edition.* Fcap. 8vo. 5*s.*

SERMONS ON NATIONAL SUBJECTS. Second Series. *Second Edition.* Fcap. 8vo. 5*s.*

ALEXANDRIA AND HER SCHOOLS. With a Preface. Crown 8vo. 5*s.*

THE LIMITS OF EXACT SCIENCE AS APPLIED TO HISTORY. An Inaugural Lecture delivered before the University of Cambridge. Crown 8vo. 2*s.*

PHAETHON; or Loose Thoughts for Loose Thinkers. *Third Edition.* Crown 8vo. 2*s.*

DAVID.—Four Sermons—David's Weakness—David's Strength—David's Anger—David's Deserts. Fcap. 8vo. cloth, 2*s.* 6*d.*

KINGSLEY. — AUSTIN ELLIOT. By HENRY KINGSLEY, Author of "Ravenshoe," &c. *New Edition.* Crown 8vo. 6*s.*

KINGSLEY. — THE RECOLLECTIONS OF GEOFFREY HAMLYN. By HENRY KINGSLEY. *Second Edition.* Crown 8vo. 6*s.*

KINGSLEY.—THE HILLYARS AND THE BURTONS: a Story of Two Families. By HENRY KINGSLEY. 3 vols. crown 8vo. cloth, 1*l.* 11*s.* 6*d.*

KINGSLEY.—RAVENSHOE. By HENRY KINGSLEY. *New Edition.* Crown 8vo. 6*s.*

KINGTON.—HISTORY of FREDERICK the SECOND, Emperor of the Romans. By T. L. KINGTON, M.A. 2 vols. demy 8vo. 32*s.*

KIRCHHOFF. — RESEARCHES on the SOLAR SPECTRUM and the SPECTRA of the CHEMICAL ELEMENTS. By G. KIRCHHOFF, of Heidelberg. Translated by HENRY E. ROSCOE, B.A. 4to. 5*s.* Also the Second Part. 4to. 5*s.* with 2 Plates.

LANCASTER.—ECLOGUES AND MONO-DRAMAS; or, a Collection of Verses. By WILLIAM LANCASTER. Extra fcap. 8vo. 4*s.* 6*d.*

LANCASTER. — PRÆTERITA: Poems. By WILLIAM LANCASTER. Extra fcap. 8vo. 4*s.* 6*d.*

LANCASTER. — STUDIES IN VERSE. By WILLIAM LANCASTER. Extra fcap. 8vo. cloth, 4*s.* 6*d.*

LATHAM. — THE CONSTRUCTION of WROUGHT-IRON BRIDGES, embracing the Practical Application of the Principles of Mechanics to Wrought-Iron Girder Work. By J. H. LATHAM, Esq. Civil Engineer. 8vo. With numerous detail Plates. *Second Edition.* [*Preparing.*

LECTURES TO LADIES ON PRACTICAL SUBJECTS. *Third Edition*, revised. Crown 8vo. 7*s*. 6*d*.

LEMON. — LEGENDS OF NUMBER NIP. By MARK LEMON. With Six Illustrations by CHARLES KEENE. Extra fcap. 5*s*.

LESLEY'S GUARDIANS: A Novel. By CECIL HOME. 3 vols. crown 8vo. 31*s*. 6*d*.

LIGHTFOOT. — ST. PAUL'S EPISTLE TO THE GALATIANS. A Revised Text, with Notes and Dissertations. By J. D. LIGHTFOOT, D.D. 8vo. cloth, 10*s*. 6*d*.

LOWELL. — FIRESIDE TRAVELS. By JAMES RUSSELL LOWELL, Author of "The Biglow Papers." Fcap. 8vo. 4*s*. 6*d*.

LUDLOW and HUGHES.— A SKETCH of the HISTORY of the UNITED STATES from Independence to Secession. By J. M. LUDLOW, Author of "British India, its Races and its History," "The Policy of the Crown towards India," &c.

To which is added, THE STRUGGLE FOR KANSAS. By THOMAS HUGHES, Author of "Tom Brown's School Days," "Tom Brown at Oxford," &c. Crown 8vo. 8*s*. 6*d*.

LUDLOW.—BRITISH INDIA ; its Races, and its History, down to 1857. By JOHN MALCOLM LUDLOW, Barrister-at-Law. 2 vols. 9*s*.

LUDLOW. — POPULAR EPICS OF THE MIDDLE AGES, OF THE NORSE-GERMAN AND CARLOVINGIAN CYCLES. By JOHN MALCOLM LUDLOW. 2 vols. fcap. 8vo. cloth, 14*s*.

LUSHINGTON.—THE ITALIAN WAR 1848-9, and the Last Italian Poet. By the late HENRY LUSHINGTON. With a Biographical Preface by G. S. VENABLES. Crown 8vo. 6*s*. 6*d*.

LYTTELTON.—THE COMUS OF MILTON rendered into Greek Verse. By LORD LYTTELTON. Extra fcap. 8vo. *Second Edition.* 5*s*.

MACKENZIE.—THE CHRISTIAN CLERGY of the FIRST TEN CENTURIES, and their Influence on European Civilization. By HENRY MAC-KENZIE, B.A. Scholar of Trinity College, Cambridge. Crown 8vo. 6*s*. 6*d*.

MACLAREN.— SERMONS PREACHED AT MANCHESTER. By ALEXANDER MACLAREN. *Second Edition.* Fcp. 8vo. 4*s*. 6*d*. A second Series in the Press.

MACLEAR.—A HISTORY OF CHRISTIAN MISSIONS DURING THE MIDDLE AGES. By G. F. MACLEAR, M.A. Crown 8vo. 10*s*. 6*d*.

MACLEAR. — THE WITNESS OF THE EUCHARIST; or, The Institution and Early Celebration of the Lord's Supper, considered as an Evidence of the Historical Truth of the Gospel Narrative and of the Atonement. Crown 8vo. 4*s*. 6*d*.

MACLEAR. — A CLASS-BOOK OF OLD TESTA-MENT HISTORY. By the Rev. G. F. MACLEAR, M.A. With Four Maps. *Second Edition.* 18mo. cloth, 4s. 6d.

MACLEAR. — A CLASS-BOOK OF NEW TESTA-MENT HISTORY, including the connexion of the Old and New Testament.

MACMILLAN. — FOOT-NOTES FROM THE PAGE OF NATURE. By the Rev. HUGH MACMILLAN, F.R.S.E. With numerous Illustrations. Fcap. 8vo. 5s.

MACMILLAN'S MAGA-ZINE. Published Monthly, price One Shilling. Volumes I.—XII. are now ready, 7s. 6d. each.

McCOSH.—The METHOD of the DIVINE GOVERN-MENT, Physical and Moral. By JAMES McCOSH, LL.D. *Eighth Edition.* 8vo. 10s. 6d.

McCOSH.—THE SUPER-NATURAL IN RELATION TO THE NATURAL. By JAMES McCOSH, LL.D. Crown 8vo. 7s. 6d.

McCOSH. — THE INTUI-TIONS OF THE MIND. By JAMES McCOSH, LL.D. *A New Edition.* 8vo. cloth, 10s. 6d.

McCOSH. — A DEFENCE OF FUNDAMENTAL TRUTH, BEING A REVIEW OF THE PHILOSOPHY OF MR. STUART MILL. By JAMES McCOSH, LL.D. Cr. 8vo. [*In the press.*

McCOY.—CONTRIBU-TIONS TO BRITISH PALÆ-ONTOLOGY; or, First Descriptions of several hundred Fossil Radiata, Articulata, Mollusca, and Pisces, from the Tertiary, Cretaceous, Oolitic, and Palæozoic Strata of Great Britain. With numerous Woodcuts. By FRED. McCOY, F.G.S. Professor of Natural History in the University of Melbourne. 8vo. 9s.

MANSFIELD. — PARA-GUAY, BRAZIL, AND THE PLATE. With a Map, and numerous Woodcuts. By CHARLES MANSFIELD, M.A. With a Sketch of his Life. By the Rev. CHARLES KINGSLEY. Crown 8vo. 12s. 6d.

MANSFIELD. — A THEORY OF SALTS. A Treatise on the Constitution of Bipolar (two membered) Chemical Compounds. By the late CHARLES BLANCHFORD MANSFIELD. Crown 8vo. cloth, price 14s.

MARRIED BENEATH HIM. By the Author of "Lost Sir Massingberd." 3 vols. crown 8vo. cloth, 1l. 11s. 6d.

MARRINER. — SERMONS PREACHED at LYME REGIS. By E. T. MARRINER, Curate. Fcap. 8vo. 4s. 6d.

MARSTON.—A LADY IN HER OWN RIGHT. By WEST-LAND MARSTON. Crown 8vo. 6s.

MARTIN.—THE STATES-MAN'S YEAR BOOK for 1866. A Statistical, Genealogical, and Historical Account of the Civilized World for the Year 1866. By FREDERICK MARTIN. Cr. 8vo. 10s. 6d.

MARTIN. — STORIES OF BANKS AND BANKERS. By FREDERICK MARTIN. Fcp. 8vo. cloth, 3*s.* 6*d.*

MARTIN.—THE LIFE OF JOHN CLARE. By FREDE-RICK MARTIN. Crown 8vo. cloth, 7*s.* 6*d.*

MASSON.—ESSAYS, BIOGRAPHICAL and CRITI-CAL; chiefly on the English Poets. By DAVID MASSON, M.A. 8vo. 12*s.* 6*d.*

MASSON.—BRITISH NOVELISTS AND THEIR STYLES; being a Critical Sketch of the History of British Prose Fiction. By DAVID MASSON, M.A. Crown 8vo. 7*s.* 6*d.*

MASSON.—LIFE of JOHN MILTON, narrated in Connexion with the Political, Ecclesiastical, and Literary History of his Time. Vol. I. with Portraits. 18*s.*

MASSON.—RECENT BRITISH PHILOSOPHY. A Review, with Criticisms. By DAVID MASSON. Crown 8vo. cloth, 7*s.* 6*d.*

MAURICE.—WORKS BY THE REV. FREDERICK DENISON MAURICE, M.A.

THE CLAIMS OF THE BIBLE AND OF SCIENCE; a Corre-spondence on some questions re-specting the Pentateuch. Crown 8vo. 4*s.* 6*d.*

DIALOGUES on FAMILY WOR-SHIP. Crown 8vo. 6*s.*

EXPOSITORY DISCOURSES on the Holy Scriptures :—

THE PATRIARCHS and LAW-GIVERS of the OLD TESTA-MENT. *Second Edition.* Crown 8vo. 6*s.*

This volume contains Discourses on the Pentateuch, Joshua, Judges, and the beginning of the First Book of Samuel.

THE PROPHETS and KINGS of the OLD TESTAMENT. *Second Edition.* Crown 8vo. 10*s.* 6*d.*

This volume contains Discourses on Samuel I. and II., Kings I. and II. Amos, Joel, Hosea, Isaiah, Micah, Nahum, Habakkuk, Jeremiah, and Ezekiel.

THE GOSPEL OF THE KING-DOM OF GOD. A Series of Lectures on the Gospel of St. Luke. Crown 8vo. 9*s.*

THE GOSPEL OF ST. JOHN ; a Series of Discourses. *Second Edition.* Crown 8vo. 10*s.* 6*d.*

THE EPISTLES OF ST. JOHN ; a Series of Lectures on Christian Ethics. Crown 8vo. 7*s.* 6*d.*

EXPOSITORY SERMONS ON THE PRAYER-BOOK :—

THE ORDINARY SERVICES. *Second Edition.* Fcap. 8vo. 5*s.* 6*d.*

THE CHURCH A FAMILY. Twelve Sermons on the Occa-sional Services. Fcap. 8vo. 4*s.* 6*d.*

LECTURES ON THE APO-CALYPSE, or, Book of the Revelation of St. John the Divine. Crown 8vo. 10*s.* 6*d.*

WHAT IS REVELATION ? A Series of Sermons on the Epiphany, to which are added Letters to a Theological Student on the Bamp-ton Lectures of Mr. MANSEL. Crown 8vo. 10*s.* 6*d.*

SEQUEL TO THE INQUIRY, "WHAT IS REVELATION?" Letters in Reply to Mr. Mansel's Examination of "Strictures on the Bampton Lectures." Crown 8vo. 6s.

LECTURES ON ECCLESIASTICAL HISTORY. 8vo. 10s. 6d.

THEOLOGICAL ESSAYS. *Second Edition.* Crown 8vo. 10s. 6d.

THE DOCTRINE OF SACRIFICE DEDUCED FROM THE SCRIPTURES. Cr. 8vo. 7s. 6d.

THE RELIGIONS OF THE WORLD, and their Relations to Christianity. *Fourth Edition.* Fcap. 8vo. 5s.

ON THE LORD'S PRAYER. *Fourth Edition.* Fcap. 8vo. 2s. 6d.

ON THE SABBATH DAY: the Character of the Warrior; and on the Interpretation of History. Fcap. 8vo. 2s. 6d.

LEARNING AND WORKING. —Six Lectures on the Foundation of Colleges for Working Men. Crown 8vo. 5s.

THE INDIAN CRISIS. Five Sermons. Crown 8vo. 2s. 6d.

LAW'S REMARKS ON THE FABLE OF THE BEES. With an Introduction by F. D. MAURICE, M.A. Fcap. 8vo. 4s. 6d.

MAYOR.—AUTOBIOGRAPHY OF MATTHEW ROBINSON. By JOHN E. B. MAYOR, M.A. Fcp. 8vo. 5s. 6d.

MAYOR. — EARLY STATUTES of ST. JOHN'S COLLEGE, CAMBRIDGE. With Notes. Royal 8vo. 18s.

MELIBŒUS IN LONDON. By JAMES PAYN, M.A. Fcap. 8vo. 2s. 6d.

MERIVALE. — SALLUST FOR SCHOOLS. By C. MERIVALE, B.D. *Second Edition.* Fcap. 8vo. 4s. 6d.
*** The Jugurtha and the Catalina may be had separately, price 2s. 6d. each.

MERIVALE.—KEATS' HYPERION Rendered into Latin Verse. By C. MERIVALE, B.D. *Second Edition.* Extra fcap. 8vo. 3s. 6d.

MISS RUSSELL'S HOBBY. A Novel. 2 vols. crown 8vo. cloth, 12s.

MOOR COTTAGE.—A Tale of Home Life. By the Author of "Little Estella." Crown 8vo. 6s.

MOORHOUSE. — SOME MODERN DIFFICULTIES respecting the FACTS of NATURE and REVELATION. By JAMES MOORHOUSE, M.A. Fcap. 8vo. 2s. 6d.

MORGAN. — A COLLECTION OF MATHEMATICAL PROBLEMS and EXAMPLES. By H. A. MORGAN, M.A. Crown 8vo. 6s. 6d.

MORSE.—WORKING FOR GOD, and other Practical Sermons. By FRANCIS MORSE, M.A. *Second Edition.* Fcap. 8vo. 5s.

MORTLOCK. — CHRISTIANITY AGREEABLE TO REASON. By the Rev. EDMUND MORTLOCK, B.D. *Second Edition.* Fcap. 8vo. 3s. 6d.

B

NAVILLE. — THE HEA-
VENLY FATHER. By ER-
NEST NAVILLE, Correspond-
ing Member of the Institute of
France, and formerly Professor of
Philosophy in the University of
Geneva. Translated by HENRY
DOWNTON, M.A. Extra fcap.
8vo. 7*s*. 6*d*.

NOEL.—BEHIND THE
VEIL, and other Poems. By the
Hon. RODEN NOEL. Fcap.
8vo. 7*s*.

NORTHERN CIRCUIT.
Brief Notes of Travel in Sweden,
Finland, and Russia. With a
Frontispiece. Crown 8vo. 5*s*.

NORTON.—THE LADY of
LA GARAYE. By the Hon.
Mrs. NORTON. With Vignette
and Frontispiece. *New Edit.*4*s*.6*d*.

O'BRIEN.—An ATTEMPT
to EXPLAIN and ESTABLISH
the DOCTRINE of JUSTIFI-
CATION BY FAITH ONLY.
By JAMES THOS. O'BRIEN,
D.D. Bishop of Ossory. *Third
Edition.* 8vo. 12*s*.

O'BRIEN.—CHARGE deli-
vered at the Visitation in 1863.
Second Edition. 8vo. 2*s*.

OLIPHANT.—AGNES
HOPETOUN'S SCHOOLS
AND HOLIDAYS. By MRS.
OLIPHANT. Royal 16mo.cloth,
gilt leaves. 3*s*. 6*d*.

OLIVER. — LESSONS IN
ELEMENTARY BOTANY.
With nearly 200 Illustrations.
By DANIEL OLIVER, F.R.S.
F.L.S. 18mo. 4*s*. 6*d*.

OPPEN.—FRENCH
READER, for the Use of Colleges and Schools. By EDWARD
A. OPPEN. Fcap. 8vo. cloth,
4*s*. 6*d*.

ORWELL.—The BISHOP'S
WALK AND THE BISHOP'S
TIMES. Poems on the Days of
Archbishop Leighton and the
Scottish Covenant. By ORWELL.
Fcap. 8vo. 5*s*.

OUR YEAR. — A Child's
Book, in Prose and Verse. By
the Author of "John Halifax,
Gentleman." Illustrated by
CLARENCE DOBELL. Royal
16mo. cloth, 3*s*. 6*d*.

PALGRAVE. — HISTORY
OF NORMANDY AND OF
ENGLAND. By Sir FRANCIS
PALGRAVE. Completing the
History to the Death of William
Rufus. Vols. I. to IV. 8vo.
each 21*s*.

PALGRAVE.—A NARRA-
TIVE OF A YEAR'S JOUR-
NEY THROUGH CENTRAL
AND EASTERN ARABIA,
1862-3. By WILLIAM GIF-
FARD PALGRAVE (late of the
Eighth Regiment Bombay N.I.).
Third Edition. 2 vols. 8vo.
cloth. 28*s*.

PALGRAVE.—ESSAYS ON
ART. By FRANCIS TUR-
NER PALGRAVE, M.A. late
Fellow of Exeter College, Oxford.
Mulready—Dyce—Holman Hunt
—Herbert—Poetry, Prose, and
Sensationalism in Art—Sculpture
in England—The Albert Cross,
&c. Extra fcap. 8vo. (Uniform
with "Arnold's Essays.")

PALGRAVE. — SONNETS
AND SONGS. By WILLIAM
SHAKESPEARE. GEM EDI-
TION. Edited by F. T PAL-
GRAVE, M.A. With Vignette
Title by JEENS, 3*s*. 6*d*.

PALMER.—THE BOOK of PRAISE : from the Best English Hymn Writers. Selected and arranged by ROUNDELL PALMER. With Vignette by WOOLNER. Pott 8vo. 4s. 6d. *Large Type Edition*, demy 8vo. 10s. 6d. ; morocco, 28s.

PARKINSON.—A TREATISE ON ELEMENTARY MECHANICS. For the Use of the Junior Classes at the University and the Higher Classes in Schools. With a Collection of Examples. By S. PARKINSON, B.D. *Third Edition*, revised. Crown 8vo. 9s. 6d.

PARKINSON.—A TREATISE ON OPTICS. By S. PARKINSON, B.D. Crown 8vo. 10s. 6d.

PATERSON. — TREATISE ON THE FISHERY LAWS of the UNITED KINGDOM, including the Laws of Angling. By JAMES PATERSON, M.A. Crown 8vo. 10s.

PATMORE.—The ANGEL IN THE HOUSE. Book I. The Betrothal.—Book II. The Espousals.—Book III. Faithful for Ever. With Tamerton Church Tower. By COVENTRY PATMORE. 2 vols. fcap. 8vo. 12s.
** A New and Cheap Edition, in 1 vol. 18mo. beautifully printed on toned paper, price 2s. 6d.

PATMORE. — THE VICTORIES OF LOVE. Fcap. 8vo. 4s. 6d.

PAULI. — PICTURES OF OLD ENGLAND. By Dr. REINHOLD PAULI. Translated by E. C. OTTE. Crown 8vo. 8s. 6d.

PEEL.—JUDAS MACCABÆUS. An Heroic Poem. By EDMUND PEEL. Fcap. 8vo. 7s. 6d.

PHEAR.—ELEMENTARY HYDROSTATICS. By J. B. PHEAR, M.A. *Third Edition*. Crown 8vo. 5s. 6d.

PHILLIMORE.—PRIVATE LAW among the ROMANS. From the Pandects. By JOHN GEORGE PHILLIMORE, Q.C. 8vo. 16s.

PHILLIPS.—LIFE on the EARTH : its Origin and Succession. By JOHN PHILLIPS, M.A. LL.D. F.R.S. With Illustrations. Crown 8vo. 6s. 6d.

PHILOLOGY.—The JOURNAL OF SACRED AND CLASSICAL PHILOLOGY. Four vols. 8vo. 12s. 6d. each.

PLATO.—The REPUBLIC OF PLATO. Translated into English, with Notes. By Two Fellows of Trinity College, Cambridge (J. Ll. Davies, M.A. and D. J. Vaughan, M.A.). With Vignette Portraits of Plato and Socrates engraved by JEENS from an Antique Gem. (Golden Treasury Series). *New Edition*, 18mo. 4s. 6d.

PLATONIC DIALOGUES, The. For English Readers. By W. WHEWELL, D.D. F.R.S. Master of Trinity College, Cambridge. Vol. I. *Second Edition*, containing *The Socratic Dialogues*, fcap. 8vo. 7s. 6d. Vol. II. containing *The Anti-Sophist Dialogues*, 6s. 6d. Vol. III. containing *The Republic*. 7s. 6d.

PLEA for a NEW ENGLISH VERSION of THE SCRIPTURES. By a Licentiate of the Church of Scotland. 8vo. 6s.

POTTER.—A VOICE from the CHURCH in AUSTRALIA: Sermons preached in Melbourne. By the Rev. ROBERT POTTER, M.A. Extra fcap. 8vo. 4s. 6d.

PRATT.—TREATISE ON ATTRACTIONS, La Place's FUNCTIONS, and the FIGURE of the EARTH. By J. H. PRATT, M.A. *Second Edition.* Crown 8vo. 6s. 6d.

PROCTER.—A HISTORY of the BOOK OF COMMON PRAYER : with a Rationale of its Offices. By FRANCIS PROCTER, M.A. *Fifth Edition,* revised and enlarged. Cr. 8vo. 10s. 6d.

PROCTER.—An ELEMENTARY HISTORY of the BOOK of COMMON PRAYER. By FRANCIS PROCTER, M.A. 18mo. 2s. 6d.

PROPERTY and INCOME.—GUIDE to the UNPROTECTED in Matters relating to Property and Income. *Second Edition.* Crown 8vo. 3s. 6d.

PUCKLE.—AN ELEMENTARY TREATISE on CONIC SECTIONS and ALGEBRAIC GEOMETRY, especially designed for the Use of Schools and Beginners. By G. HALE PUCKLE, M.A. *Second Edition.* Crown 8vo. 7s. 6d.

RAMSAY. — THE CATECHISER'S MANUAL ; or, the Church Catechism illustrated and explained, for the Use of Clergymen, Schoolmasters, and Teachers. By ARTHUR RAMSAY, M.A. *Second Edition.* 18mo. 1s. 6d.

RAWLINSON.—ELEMENTARY STATICS. By G. RAWLINSON, M.A. Edited by EDWARD STURGES, M.A. Crown 8vo. 4s. 6d.

RAYS of SUNLIGHT for DARK DAYS. A Book of Selections for the Suffering. With a Preface by C. J. VAUGHAN, D.D. 18mo. *New Edition.* 3s. 6d. ; morocco, old style, 9s.

REYNOLDS.—A SYSTEM OF MEDICINE. To be completed in Three Volumes, 8vo. Edited by J. RUSSELL REYNOLDS, M.D. F.R.C.P. London. The First Volume will contain : — PART I. — GENERAL DISEASES, or Affections of the Whole System. § I.—Those determined by agents operating from without, such as the exanthemata, malarial diseases, and their allies. § II.—Those determined by conditions existing within the body, such as Gout, Rheumatism, Rickets, &c. PART II.—LOCAL DISEASES, or Affections of Particular Systems. § I.—Diseases of the Skin. [*In the Press.*

REYNOLDS.—NOTES OF THE CHRISTIAN LIFE. A Selection of Sermons by HENRY ROBERT REYNOLDS, B.A. President of Cheshunt College, and Fellow of University College, London. Crown 8vo. cloth, price 7s. 6d.

ROBERTS.—DISCUSSIONS ON THE GOSPELS. By REV. ALEXANDER ROBERTS, D.D. *Second Edition,* revised and enlarged. 8vo. cloth, 16s.

ROBY. — AN ELEMEN-
TARY LATIN GRAMMAR.
By H. J. ROBY, M.A. 18mo.
2s. 6d.

ROBY.—STORY OF A
HOUSEHOLD, and Other
Poems. By MARY K. ROBY.
Fcap. 8vo. 5s.

ROMANIS.—SERMONS
PREACHED at ST. MARY'S,
READING. By WILLIAM
ROMANIS, M.A. *First Series.*
Fcap. 8vo. 6s. Also, *Second
Series.* 6s.

ROSSETTI.—GOBLIN
MARKET, and other Poems.
By CHRISTINA ROSSETTI.
With Two Designs by D. G.
ROSSETTI. *Second Edition.* Fcap.
8vo. 5s.

ROSSETTI. — THE
PRINCE'S PROGRESS, and
other Poems. By CHRISTINA
ROSSETTI. With Two Designs
by D. G. ROSSETTI.

ROSSETTI. — DANTE'S
COMEDY: *The Hell.* Translated into Literal Blank Verse.
By W. M. ROSSETTI. Fcap.
8vo. cloth. 5s.

ROUTH.—TREATISE ON
DYNAMICS OF RIGID BO-
DIES. With Numerous Exam-
ples. By E. J. ROUTH, M.A.
Crown 8vo. 10s. 6d.

ROWSELL.—The ENGLISH
UNIVERSITIES AND THE
ENGLISH POOR. Sermons
preached before the University of
Cambridge. By T. J. ROW-
SELL, M.A. Fcap. 8vo. 2s.

ROWSELL. — MAN'S
LABOUR and GOD'S HAR-
VEST. Sermons preached before the University of Cambridge
in Lent, 1861. Fcap. 8vo.
3s.

RUFFINI. — VINCENZO ;
or, SUNKEN ROCKS. By
JOHN RUFFINI. Three vols.
crown 8vo. 31s. 6d.

RUTH and her FRIENDS.
A Story for Girls. With a Frontispiece. *Fourth Edition.* Royal
16mo. 3s. 6d.

SCOURING of the WHITE
HORSE ; or, the Long Vacation
Ramble of a London Clerk. By
the Author of "Tom Brown's
School Days." Illustrated by
DOYLE. *Eighth Thousand.* Imp.
16mo. 8s. 6d.

SELWYN.— THE WORK
of CHRIST in the WORLD.
By G. A. SELWYN, D.D.
Third Edition. Crown 8vo.
2s.

SHAKESPEARE.—THE
WORKS OF WILLIAM
SHAKESPEARE. Edited by
WM. GEORGE CLARK, M.A.
and W. ALDIS WRIGHT,
M.A. Vols. 1 to 8, 8vo. 10s. 6d.
each. To be completed in Nine
Volumes.

SHAKESPEARE.—THE
COMPLETE WORKS OF
WILLIAM SHAKESPEARE.
The *Globe Edition.* Edited by
W. G. CLARK and W. A.
WRIGHT. Fifty-first Thousand.
Royal Fcap. 3s. 6d.

SHAKESPEARE. — SON-
NETS AND SONGS. By
WILLIAM SHAKESPEARE.
Edited by FRANCIS TURNER
PALGRAVE, M.A. The GEM
EDITION. With Vignette Title,
price 3s. 6d.

SHAKESPEARE'S TEM-
PEST. With Glossarial and Ex-
planatory Notes. By the Rev. J.
M. JEPHSON. 18mo. 3s. 6d.

SHAIRP. — KILMAHOE:
and other Poems. By J. CAMP-
BELL SHAIRP. Fcap. 8vo. 5s

SHIRLEY.—ELIJAH; Four
University Sermons. I. Samaria.
II. Carmel.—III. Kishon.—IV.
Horeb. By W. W. SHIRLEY,
D.D. Fcap. 8vo. 2s. 6d.

SIMEON.—STRAY NOTES
ON FISHING AND ON
NATURAL HISTORY. By
CORNWALL SIMEON. Cr.
8vo. 7s. 6d.

SIMPSON.—AN EPITOME
OF THE HISTORY OF THE
CHRISTIAN CHURCH. By
WILLIAM SIMPSON, M.A.
Fourth Edition. Fcp. 8vo. 3s. 6d.

SKETCHES FROM CAM-
BRIDGE. By A DON. Crown
8vo. cloth, 3s. 6d.

SMITH.—A LIFE DRAMA,
and other Poems. By ALEX-
ANDER SMITH. Fcap. 8vo.
2s. 6d.

SMITH. — CITY POEMS.
By ALEXANDER SMITH,
Fcap. 8vo. 5s.

SMITH.—EDWIN OF
DEIRA. Second Edition. By
ALEXANDER SMITH. Fcap.
8vo. 5s.

SMITH.—A LETTER TO
A WHIG MEMBER of the
SOUTHERN INDEPEN-
DENCE ASSOCIATION. By
GOLDWIN SMITH. Extra
fcap. 8vo. 2s.

SMITH. — ARITHMETIC
AND ALGEBRA. By BAR-
NARD SMITH, M.A. Ninth
Edition. Cr. 8vo. cloth, 10s. 6d.

SMITH. — ARITHMETIC
for the USE of SCHOOLS.
New Edition. Crown 8vo. 4s. 6d.

SMITH.—A KEY to the
ARITHMETIC for SCHOOLS.
Second Edition. Crown 8vo.
8s. 6d.

SMITH.—EXERCISES IN
ARITHMETIC. By BAR-
NARD SMITH. With Answers.
Crown 8vo. limp cloth, 2s. 6d.
Or sold separately, as follows :—
Part I. 1s. Part II. 1s. Answers,
6d.

SMITH.—SCHOOL CLASS
BOOK of ARITHMETIC. By
BARNARD SMITH. 18mo.
cloth, 3s. Or sold separately,
Parts I. and II. 10d. each, Part
III. 1s.

SMITH. — KEYS TO
SCHOOL CLASS BOOK OF
ARITHMETIC. By BAR-
NARD SMITH. Complete in
One Volume, 18mo. 6s. 6d.; or
Parts I. II. and III. 2s. 6d. each.

SMITH.—A ONE SHILLING
BOOK of ARITHMETIC for
NATIONAL and ELEMEN-
TARY SCHOOLS. By BAR-
NARD SMITH. 18mo. cloth.

SNOWBALL. — THE ELE-
MENTS of PLANE and SPHE-
RICAL TRIGONOMETRY.
By J. C. SNOWBALL, M.A.
Tenth Edition. Crown 8vo. 7s. 6d.

SPRING SONGS. — By a WEST HIGHLANDER. With a Vignette Illustration by GOURLAY STEELE. Fcap. 8vo. 1s. 6d.

STEPHEN. — GENERAL VIEW of the CRIMINAL LAW of ENGLAND. By J. FITZJAMES STEPHEN. 8vo. 18s.

STORY.—MEMOIR of the Rev. ROBERT STORY. By R. H. STORY. Crown 8vo. 7s. 6d.

STRICKLAND.—ON COTTAGE CONSTRUCTION and DESIGN. By C. W. STRICKLAND. With Specifications and Plans. 8vo. 7s. 6d.

SWAINSON. — A HANDBOOK to BUTLER'S ANALOGY. By C. A. SWAINSON, D.D. Crown 8vo. 1s. 6d.

SWAINSON.—The CREEDS of the CHURCH in their RELATIONS to HOLY SCRIPTURE and the CONSCIENCE of the CHRISTIAN. 8vo. cloth, 9s.

SWAINSON.—The AUTHORITY of the NEW TESTAMENT, and other Lectures, delivered before the University of Cambridge. 8vo. cloth, 12s.

TACITUS.—The HISTORY of TACITUS translated into ENGLISH. By A. J. CHURCH, M.A., and W. J. BRODRIBB, M.A. With a Map and Notes. 8vo. 10s. 6d.

TAIT AND STEELE.—A TREATISE ON DYNAMICS, with numerous Examples. By P. G. TAIT and W. J. STEELE. Second Edition. Crown 8vo. 10s. 6d.

TAYLOR.—WORDS AND PLACES; or, Etymological Illustrations of History, Ethnology, and Geography. By the Rev. ISAAC TAYLOR. Second Edition. Crown 8vo. 12s. 6d.

TAYLOR.—THE RESTORATION OF BELIEF. New and Revised Edition. By ISAAC TAYLOR, Esq. Crown 8vo. 8s. 6d.

TAYLOR.—BALLADS AND SONGS OF BRITTANY. By TOM TAYLOR. With Illustrations by TISSOT, MILLAIS, TENNIEL, KEENE, and H. K. BROWNE. Small 4to. cloth gilt, 12s.

TAYLOR. — GEOMETRICAL CONICS. By C. TAYLOR, B.A. Crown 8vo. 7s. 6d.

TEMPLE. — SERMONS PREACHED in the CHAPEL of RUGBY SCHOOL. By F. TEMPLE, D.D. 8vo. 10s. 6d.

THORPE.— DIPLOMATARIUM ANGLICUM ÆVI SAXONICI. A Collection of ENGLISH CHARTERS, from the Reign of King Æthelberht of Kent, A.D. DC.V. to that of William the Conqueror. With a Translation of the Anglo-Saxon. By BENJAMIN THORPE, Member of the Royal Academy of Sciences, Munich. 8vo. cloth, price 21s.

THRING.—A CONSTRUING BOOK. Compiled by EDWARD THRING, M.A. Fcap. 8vo. 2s. 6d.

THRING.—A LATIN GRA-
DUAL. A First Latin Construing Book for Beginners. Fcap.
8vo. 2s. 6d.

THRING.—THE ELE-
MENTS of GRAMMAR taught
in ENGLISH. *Third Edition.*
18mo. 2s.

THRING.—THE CHILD'S
GRAMMAR. *A New Edition.*
18mo. 1s.

THRING. — SERMONS
DELIVERED at UPPINGHAM
SCHOOL. Crown 8vo. 5s.

THRING.—SCHOOL
SONGS. With the Music ar-
ranged for four Voices. Edited
by the Rev. EDWD. THRING,
M.A. and H. RICCIUS. Small
folio, 7s. 6d.

THRING. — EDUCATION
and SCHOOL. By the Rev.
EDWARD THRING, M.A.
Crown 8vo. 6s. 6d.

THRUPP.—The SONG of
SONGS. A New Translation,
with a Commentary and an Introduction. By the Rev. J. F.
THRUPP. Crown 8vo. 7s. 6d.

THRUPP. — ANTIENT
JERUSALEM : a New Investi-
gation into the History, Topo-
graphy, and Plan of the City,
Environs, and Temple. With
Map and Plans. 8vo. 15s.

THRUPP. — INTRODUC-
TION to the STUDY and USE
of the PSALMS. 2 vols. 21s.

THRUPP—PSALMS AND
HYMNS for PUBLIC WOR-
SHIP. Selected and Edited by
the Rev. J. F. THRUPP, M.A.
18mo. 2s. common paper, 1s. 4d.

TOCQUEVILLE. — ME-
MOIR, LETTERS, and RE-
MAINS of ALEXIS DE TOC-
QUEVILLE. Translated from
the French by the Translator of
"Napoleon's Correspondence with
King Joseph." With Numerous
additions, 2 vols. crown 8vo.
21s.

TODD.—THE BOOKS OF
THE VAUDOIS. The Walden-
sian Manuscripts preserved in the
Library of Trinity College, Dub-
lin, with an Appendix by JAMES
HENTHORN TODD, D.D.
Crown 8vo. cloth, 6s.

TODHUNTER. — WORKS
by ISAAC TODHUNTER,
M.A. F.R.S.

EUCLID FOR COLLEGES
AND SCHOOLS. *New Edition.*
18mo. 3s. 6d.

ALGEBRA FOR BEGINNERS.
With numerous Examples. 18mo.
2s. 6d.

A TREATISE ON THE DIF-
FERENTIAL CALCULUS.
With numerous Examples. *Fourth
Edition.* Crown 8vo. 10s 6d.

A TREATISE ON THE IN-
TEGRAL CALCULUS. *Second
Edition.* With numerous Exam-
ples. Crown 8vo. 10s. 6d.

A TREATISE ON ANALYTI-
CAL STATICS. *Second Edition.*
Crown 8vo. 10s. 6d.

A TREATISE ON CONIC SEC
TIONS. *Third Edition.* Crown
8vo. 7s. 6d.

ALGEBRA FOR THE USE OF
COLLEGES AND SCHOOLS.
Third Edition. Crown 8vo.
7s. 6d.

PLANE TRIGONOMETRY for COLLEGES and SCHOOLS. *Third Edition.* Crown 8vo. 5*s.*

A TREATISE ON SPHERICAL TRIGONOMETRY for the USE of COLLEGES and SCHOOLS. *Second Edition.* Crown 8vo. 4*s.*6*d.*

CRITICAL HISTORY OF THE PROGRESS of the CALCULUS of VARIATIONS during the NINETEENTH CENTURY. 8vo. 12*s.*

EXAMPLES OF ANALYTICAL GEOMETRY of THREE DIMENSIONS. *Second Edition.* Crown 8vo. 4*s.*

A TREATISE on the THEORY of EQUATIONS. Crown 8vo. cloth, 7*s.* 6*d.*

MATHEMATICAL THEORY OF PROBABILITY. 8vo. cloth, 18*s.*

TOM BROWN'S SCHOOL DAYS. By an OLD BOY. 31*st* *Thousand.* Fcap. 8vo. 5*s.* (*People's Edition,* 2*s.*)

TOM BROWN at OXFORD. By the Author of "Tom Brown's School Days." *New Edition.* Crown 8vo. 6*s.*

TRACTS FOR PRIESTS and PEOPLE. By VARIOUS WRITERS.

THE FIRST SERIES, Crown 8vo. 8*s.*

THE SECOND SERIES, Crown 8vo. 8*s.*

The whole Series of Fifteen Tracts may be had separately, price One Shilling each.

TRENCH. — WORKS BY R. CHENEVIX TRENCH, D.D. Archbishop of Dublin.

NOTES ON THE PARABLES OF OUR LORD. *Ninth Edition.* 8vo. 12*s.*

NOTES ON THE MIRACLES OF OUR LORD. *Seventh Edition.* 8vo. 12*s.*

SYNONYMS OF THE NEW TESTAMENT. *New Edition.* 1 vol. 8vo. cloth, 10*s.* 6*d.*

ON THE STUDY OF WORDS. *Eleventh Edition.* Fcap. 4*s.*

ENGLISH PAST AND PRESENT. *Fifth Edition.* Fcap. 8vo. 4*s.*

PROVERBS and their LESSONS. *Fifth Edition.* Fcap. 8vo. 3*s.*

SELECT GLOSSARY OF ENGLISH WORDS used Formerly in SENSES different from the PRESENT. *Third Edition.* 4*s.*

ON SOME DEFICIENCIES IN our ENGLISH DICTIONARIES. *Second Edition.* 8vo. 3*s.*

SERMONS PREACHED IN WESTMINSTER ABBEY. *Second Edition.* 8vo. 10*s.* 6*d.*

THE FITNESS OF HOLY SCRIPTURE for UNFOLDING the SPIRITUAL LIFE of MAN: Christ the Desire of all Nations; or, the Unconscious Prophecies of Heathendom. Hulsean Lectures. Fcap. 8vo. *Fourth Edition.* 5*s.*

ON THE AUTHORIZED VERSION of the NEW TESTAMENT. *Second Edition.* 7*s.*

POEMS. 7*s.* 6*d.*

JUSTIN MARTYR and OTHER POEMS. *Fifth Edition.* Fcap. 8vo. 6*s.*

ARCHBISHOP TRENCH'S WORKS (*continued*)—

GUSTAVUS ADOLPHUS. SOCIAL ASPECTS OF THE THIRTY YEARS' WAR. Fcp. 8vo. cloth, price 2s. 6d.

POEMS. Collected and Arranged Anew. Fcp. 8vo. 7s. 6d.

POEMS FROM EASTERN SOURCES, GENOVEVA, and other Poems. *Second Edition.* 5s. 6d.

ELEGIAC POEMS. *Third Edition.* 2s. 6d.

CALDERON'S LIFE'S A DREAM: the Great Theatre of the World. With an Essay on his Life and Genius. 4s. 6d.

REMAINS OF THE LATE MRS. RICHARD TRENCH. Being Selections from her Journals, Letters, and other Papers. *Second Edition.* With Portrait, 8vo. 15s.

COMMENTARY ON THE EPISTLES TO THE SEVEN CHURCHES IN ASIA. *Second Edition.* 8s. 6d.

SACRED LATIN POETRY. Chiefly Lyrical. Selected and Arranged for Use. *Second Edition.* Corrected and Improved. Fcap. 8vo. 7s.

TRENCH—BRIEF NOTES on the GREEK of the NEW TESTAMENT (for English Readers). By the Rev. FRANCIS TRENCH, M.A. Crown 8vo. cloth, 6s.

TRENCH.—FOUR ASSIZE SERMONS, Preached at York and Leeds. By the Rev. FRANCIS TRENCH, M.A. Crown 8vo cloth, 2s. 6d.

TREVELYAN.—THE COMPETITION WALLAH. By G. O. TREVELYAN. *New Edition.* Cr. 8vo. 6s.

TREVELYAN.—CAWNPORE. By G. O. TREVELYAN. Illustrated with Plan. *Second Edition.* Crown 8vo. 6s.

TUDOR.—THE DECALOGUE VIEWED AS THE CHRISTIAN'S LAW, with Special Reference to the Questions and Wants of the Times. By the Rev. RICH. TUDOR, B.A. Crown 8vo. 10s. 6d.

TULLOCH.—The CHRIST OF THE GOSPELS AND THE CHRIST OF MODERN CRITICISM. Lectures on M. RENAN'S "Vie de Jésus." By JOHN TULLOCH, D.D. Principal of the College of St. Mary, in the University of St. Andrew. Extra fcap. 8vo. 4s. 6d.

TURNER.—SONNETS by the Rev. CHARLES TENNYSON TURNER. Dedicated to his brother, the Poet Laureate. Fcap. 8vo. 4s. 6d.

TYRWHITT.—THE SCHOOLING OF LIFE. By R. St. JOHN TYRWHITT, M.A. Vicar of St. Mary Magdalen, Oxford. Fcap. 8vo. 3s. 6d.

VACATION TOURISTS; and Notes of Travel in 1861. Edited by F. GALTON, F.R.S. With Ten Maps illustrating the Routes. 8vo. 14s.

VACATION TOURISTS; and Notes of Travel in 1862 and 3. Edited by FRANCIS GALTON, F.R.S. 8vo. 16s.

VAUGHAN. — SERMONS PREACHED in ST. JOHN'S CHURCH, LEICESTER, during the Years 1855 and 1856. By DAVID J. VAUGHAN, M.A. Vicar of St. Martin's, Leicester. Crown 8vo. 5s. 6d.

VAUGHAN. — SERMONS ON THE RESURRECTION. With a Preface. By D. J. VAUGHAN, M.A. Fcap. 8vo. 3s.

VAUGHAN.—THREE SERMONS ON THE ATONEMENT. By D. J. VAUGHAN, M.A. 1s. 6d.

VAUGHAN. — SERMONS ON SACRIFICE AND PROPITIATION. By D. J. VAUGHAN, M.A. 2s. 6d.

VAUGHAN.—CHRISTIAN EVIDENCES and the BIBLE. By DAVID J. VAUGHAN, M.A. *New Edition.* Revised and enlarged. Fcap. 8vo. cloth, price 5s. 6d.

VAUGHAN.—WORKS BY CHARLES J. VAUGHAN, D.D. Vicar of Doncaster :—

NOTES FOR LECTURES ON CONFIRMATION. With suitable Prayers. *Sixth Edition.* 1s. 6d.

LECTURES on the EPISTLE to the PHILIPPIANS. *Second Edition.* 7s. 6d.

LECTURES on the REVELATION of ST. JOHN. 2 vols. crown 8vo. 15s. *Second Edition.* 15s.

EPIPHANY, LENT, AND EASTER. A Selection of Expository Sermons. *Second Edition.* Crown 8vo. 10s. 6d.

THE BOOK AND THE LIFE: and other Sermons Preached before the University of Cambridge. *Second Edition.* Fcap. 8vo. 4s. 6d.

MEMORIALS OF HARROW SUNDAYS. A Selection of Sermons preached in Harrow School Chapel. With a View of the Chapel. *Fourth Edition.* Cr. 8vo. 10s. 6d.

ST. PAUL'S EPISTLE TO THE ROMANS. The Greek Text with English Notes. *Second Edition.* Crown 8vo. red leaves, 5s.

REVISION OF THE LITURGY. Four Discourses. With an Introduction. I. ABSOLUTISM. II. REGENERATION. III. ATHANASIAN CREED. IV. BURIAL SERVICE. V. HOLY ORDERS. *Second Edit.* Cr. 8vo. red leaves, 4s. 6d.

LESSONS OF LIFE AND GODLINESS. A Selection of Sermons Preached in the Parish Church of Doncaster. *Third Edition.* Fcap. 8vo. 4s. 6d.

WORDS from the GOSPELS. A Second Selection of Sermons Preached in the Parish Church of Doncaster. *Second Edition.* Fcap. 8vo. 4s. 6d.

THE EPISTLES of ST. PAUL. For English Readers. Part I. containing the First Epistle to the Thessalonians. 8vo. 1s. 6d. Each Epistle will be published separately.

THE CHURCH OF THE FIRST DAYS :—
Series I. The Church of Jerusalem.
,, II. The Church of the Gentiles.
,, III. The Church of the World.
Fcap. 8vo. cloth, 4s. 6d. each.

LIFE'S WORK AND GOD'S DISCIPLINE. Three Sermons. Fcap. 8vo. cloth, 2s. 6d.

VAUGHAN.—MEMOIR of ROBERT A. VAUGHAN, Author of "Hours with the Mystics." By ROB. VAUGHAN, D.D. *Second Edition.* Revised and enlarged. Extra fcap. 8vo. 5s.

VILLAGE SERMONS BY A NORTHAMPTONSHIRE RECTOR. With a Preface on the Inspiration of Holy Scripture. Crown 8vo. 6s.

VIRGIL. — THE ÆNEID Translated into English Blank VERSE. By JOHN MILLER. Crown 8vo. 10s. 6d.

VOLUNTEER'S SCRAP BOOK. By the Author of "The Cambridge Scrap Book." Crown 4to. 7s. 6d.

WAGNER.—MEMOIR OF THE REV. GEORGE WAGNER, late of St. Stephen's, Brighton. By J. N. SIMPKINSON, M.A. *Third and Cheaper Edition.* 5s.

WARREN.—AN ESSAY on GREEK FEDERAL COINAGE. By the Hon. J. LEICESTER WARREN, M.A. 8vo. 2s. 6d.

WESTCOTT. — HISTORY of the CANON of the NEW TESTAMENT during the First Four Centuries. By BROOKE FOSS WESTCOTT, M.A. Cr. 8vo. *New Edition.* Revised. [*In the press.*

WESTCOTT. — CHARACTERISTICS of the GOSPEL MIRACLES. Sermons Preached before the University of Cambridge. *With Notes.* By B. F. WESTCOTT, M.A. Crown 8vo. 4s. 6d.

WESTCOTT. — INTRODUCTION TO THE STUDY OF THE FOUR GOSPELS. By B. F. WESTCOTT, M.A. Crown 8vo. 10s. 6d.

WESTCOTT.—The BIBLE in the CHURCH. A Popular Account of the Collection and Reception of the Holy Scriptures in the Christian Churches. By B. F. WESTCOTT, M.A. 18mo. 4s. 6d.

WESTMINSTER PLAYS.— Sive Prologi et Epilogi ad Fabulas in Sti Petri Colleg: actas qui Exstabant collecti et justa quoad licuit annorum serie ordinati, quibus accedit Declamationum qui vocantur et Epigrammatum delectus cur. F. MURE, A.M., H. BULL, A.M., CAROLO B. SCOTT, B.D. 8vo. 12s. 6d.

WILSON.—COUNSELS OF AN INVALID : Letters on Religious Subjects. By GEORGE WILSON, M.D. With Vignette Portrait. Fcap. 8vo. 4s. 6d.

WILSON.—RELIGIO CHEMICI. By GEORGE WILSON, M.D. With a Vignette beautifully engraved after a Design by NOEL PATON. Crown 8vo. 8s. 6d.

WILSON — MEMOIR OF GEORGE WILSON, M.D. F.R.S.E. Regius Professor of Technology in the University of Edinburgh. By his Sister. Third Thousand. 8vo. with Portrait. 10s. 6d.

WILSON. — THE FIVE GATEWAYS OF KNOWLEDGE. By GEORGE WILSON, M.D. *New Edit.* Fcap. 8vo. 2s. 6d. or in Paper Covers. 1s.

WILSON.—The PROGRESS of the TELEGRAPH. Fcap. 8vo. 1s.

WILSON.—PREHISTORIC ANNALS of SCOTLAND. By DANIEL WILSON, LL.D. Author of "Prehistoric Man," &c. 2 vols. demy 8vo. *New Edition.* With numerous Illustrations. 36s.

WILSON.—PREHISTORIC MAN. By DANIEL WILSON, I.L.D. *New Edition.* Revised and partly re-written, with numerous Illustrations. 1 vol. 8vo. 21s.

WILSON. — A TREATISE ON DYNAMICS. By W. P. WILSON, M.A. 8vo. 9s. 6d.

WILTON.—THE NEGEB; or, "South Country" of Scripture. By the Rev. E. WILTON, M.A. Crown 8vo. 7s. 6d.

WOLFE.—ONE HUNDRED AND FIFTY ORIGINAL PSALM AND HYMN TUNES. For Four Voices. By ARTHUR WOLFE, M.A. 10s. 6d.

WOLFE. — HYMNS FOR PUBLIC WORSHIP. Selected and arranged by ARTHUR WOLFE, M.A. 18mo. 2s. Common Paper Edition, 1s. or twenty-five for 1l.

WOLFE. — HYMNS FOR PRIVATE USE.—Selected and arranged by ARTHUR WOLFE, M.A. 18mo. 2s.

WOODFORD.—CHRISTIAN SANCTITY. By JAMES RUSSELL WOODFORD, M.A. Fcap. 8vo. cloth. 3s.

WOODWARD. — ESSAYS, THOUGHTS and REFLECTIONS, and LETTERS. By the Rev. HENRY WOODWARD. Edited by his Son. *Fifth Edition.* 8vo. cloth. 10s. 6d.

WOODWARD.—THE SHUNAMITE. By the Rev. HENRY WOODWARD, M.A. Edited by his Son, THOMAS WOODWARD, M.A. Dean of Down. *Second Edition.* Crown 8vo. cloth. 10s. 6d.

WOOLLEY. — LECTURES DELIVERED IN AUSTRALIA. By JOHN WOOLLEY, D.C.L. Crown 8vo. 8s. 6d.

WOOLNER. — MY BEAUTIFUL LADY. By THOMAS WOOLNER. With a Vignette by ARTHUR HUGHES. *Third Edition.* Fcap. 8vo. 5s.

WORDS FROM THE POETS. Selected by the Editor of "Rays of Sunlight." 18mo. extra cloth gilt, 3s. 6d.

WORSHIP (THE) OF GOD AND FELLOWSHIP AMONG MEN—Sermons on Public Worship. By MAURICE and Others. Fcap. 8vo. cloth. 3s. 6d.

WORSLEY. — CHRISTIAN DRIFT OF CAMBRIDGE WORK. Eight Lectures. By T. WORSLEY, D.D. Crown. 8vo. cloth, 6s.